Risk Analysis

To

*Hélène
Anthime
Anne
Ambroise*

Risk Analysis

Socio-technical and Industrial Systems

Jean-Marie Flaus

Series Editor
Jean-Paul Bourrières

iSTE   WILEY

First published 2013 in Great Britain and the United States by ISTE Ltd and John Wiley & Sons, Inc.

Apart from any fair dealing for the purposes of research or private study, or criticism or review, as permitted under the Copyright, Designs and Patents Act 1988, this publication may only be reproduced, stored or transmitted, in any form or by any means, with the prior permission in writing of the publishers, or in the case of reprographic reproduction in accordance with the terms and licenses issued by the CLA. Enquiries concerning reproduction outside these terms should be sent to the publishers at the undermentioned address:

ISTE Ltd
27-37 St George's Road
London SW19 4EU
UK

www.iste.co.uk

John Wiley & Sons, Inc.
111 River Street
Hoboken, NJ 07030
USA

www.wiley.com

© ISTE Ltd 2013

The rights of Jean-Marie Flaus to be identified as the author of this work have been asserted by him in accordance with the Copyright, Designs and Patents Act 1988.

Library of Congress Control Number: 2013940050

British Library Cataloguing-in-Publication Data
A CIP record for this book is available from the British Library
ISBN: 978-1-84821-492-7

Printed and bound in Great Britain by CPI Group (UK) Ltd., Croydon, Surrey CR0 4YY

Table of Contents

Foreword xiii

PART 1. GENERAL CONCEPTS AND PRINCIPLES 1

Chapter 1. Introduction 3
   1.1. What is risk management? 3
   1.2. Nature of risks 4
   1.3. Evolution of risk management 6
   1.4. Aims of this book 12

Chapter 2. Basic Notions 13
   2.1. Formalization of the notion of risk 13
   2.2. Hazard and sources of hazard 16
   2.3. Stakes and targets 17
   2.4. Vulnerability and resilience 18
   2.5. Undesirable events and scenarios 18
   2.6. Accidents and incidents 20
   2.7. Safety 20
   2.8. Likelihood, probability and frequency 21
   2.9. Severity and intensity 22
   2.10. Criticality 23
   2.11. Reducing risk: prevention, protection and barriers 23
   2.12. Risk analysis and risk management 25
   2.13. Inductive and deductive approaches 26
   2.14. Known risks and emerging risks 27
   2.15. Individual and societal risks 27

2.16. Acceptable risk 28
2.17. The ALARP and ALARA principles 29
2.18. Risk maps 31

Chapter 3. Principles of Risk Analysis Methods 33

3.1. Introduction 33
3.2. Categories of targets and damages 35
3.3. Classification of sources and undesirable events 36
    3.3.1. General points 36
    3.3.2. Case of occupational risks 38
    3.3.3. Case of major industrial risks 39
3.4. Causes of technical origin 40
    3.4.1. Material failures 41
    3.4.2. Failures in software and information systems 44
    3.4.3. Failures linked to fluids and products 45
3.5. Causes linked to the natural or manmade environment . . . 46
3.6. Human and organizational factors 46
    3.6.1. Reason's analysis of the human factor 48
    3.6.2. Tripod classification of organizational failures . . 51

Chapter 4. The Risk Management Process (ISO31000) 53

4.1. Presentation 53
4.2. ISO31000 standard 55
    4.2.1. Basic principles 55
    4.2.2. The organizational framework 56
4.3. Implementation: the risk management process 61
    4.3.1. Establishing the context 61
    4.3.2. Risk assessment 65
    4.3.3. Treatment of risk 66
    4.3.4. Communication and consultation 67
    4.3.5. Monitoring and review 68
    4.3.6. Risk evaluation methods 68

PART 2. KNOWLEDGE REPRESENTATION 71

Chapter 5. Modeling Risk 73

5.1. Introduction 73
5.2. Degradation flow models 74

    5.2.1. Source–target model 74
    5.2.2. Reason's model 75
    5.2.3. From source–target to causal modeling 77
  5.3. Causal modeling 77
    5.3.1. Fishbone cause–effect diagram 78
    5.3.2. Causal trees 79
    5.3.3. Fault tree 80
    5.3.4. Consequence or event trees 82
    5.3.5. Bow-tie diagram 83
    5.3.6. Scenario 84
    5.3.7. Bayesian networks 84
  5.4. Modeling dynamic aspects 87
    5.4.1. Markov model 87
    5.4.2. Dynamic fault tree 89
  5.5. Summary 90

Chapter 6. Measuring the Importance of a Risk 93

  6.1. Introduction 93
  6.2. Assessing likelihood 96
    6.2.1. Presentation 96
    6.2.2. Quantitative scale 97
    6.2.3. Qualitative scale 100
    6.2.4. Determining likelihood values 101
  6.3. Assessment of severity 102
    6.3.1. Presentation 102
    6.3.2. Quantitative indicators 104
    6.3.3. Qualitative indicators 104
    6.3.4. Determining a severity value 107
  6.4. Risk assessment 109
    6.4.1. Criticality 109
    6.4.2. Risk matrices 110
    6.4.3. Acceptability of a risk 112
  6.5. Application to the case of occupational risks 113
    6.5.1. Probability assessment 113
    6.5.2. Severity assessment 116
    6.5.3. Risk matrices 116
  6.6. Application to the case of industrial risks 118
    6.6.1. Probability assessment 118

6.6.2. Severity assessment 118
6.6.3. Risk matrices 120

Chapter 7. Modeling of Systems for Risk Analysis 123

7.1. Introduction 123
   7.1.1. Why model a system? 123
   7.1.2. The modeling process 124
7.2. Systemic or process modeling 126
   7.2.1. Principle 126
   7.2.2. Hierarchical breakdown 128
7.3. Functional modeling 128
   7.3.1. Identifying functions 129
   7.3.2. IDEF0 (SADT) representation 131
7.4. Structural modeling 131
7.5. Structuro-functional modeling 134
7.6. Modeling the behavior of a system 137
7.7. Modeling human tasks 140
   7.7.1. Hierarchical task analysis (HTA) 141
   7.7.2. Modeling using a decision/action flow diagram 144
   7.7.3. Event tree modeling 144
7.8. Choosing an approach 145
7.9. Relationship between the system model and the risk model . . . 146

PART 3. RISK ANALYSIS METHODS 151

Chapter 8. Preliminary Hazard Analysis 153

8.1. Introduction 153
8.2. Implementation of the method 155
   8.2.1. Definition of context, information gathering and
   representation of the installation 156
   8.2.2. Identification of hazards and undesirable events 158
   8.2.3. Analysis of hazardous situations, consequences and
   existing barriers 159
   8.2.4. Assessment of severity and frequency or likelihood 163
   8.2.5. Proposing new barriers 163
   8.2.6. Limitations 164
8.3. Model-driven PHA 165
8.4. Variations of PHA 166

        8.4.1. Different forms of results tables 166
        8.4.2. PHA in the chemical industry 167
    8.5. Examples of application 169
        8.5.1. Desk lamp 169
        8.5.2. Chemical reactor 171
        8.5.3. Automobile repair garage 172
        8.5.4. Medication circuit 173
    8.6. Summary 175

Chapter 9. Failure Mode and Effects Analysis 179

    9.1. Introduction 179
    9.2. Key concepts 181
        9.2.1. Basic definitions 181
        9.2.2. Causes of failure 181
        9.2.3. The effects of a failure 183
        9.2.4. Frequency or probability of a failure 184
        9.2.5. The severity of a failure 185
        9.2.6. Detection of a failure 185
        9.2.7. Criticality of a failure and RPN 186
    9.3. Implementation of the method 187
        9.3.1. Analysis preparation 189
        9.3.2. System modeling 189
        9.3.3. Application of the analysis procedure 190
        9.3.4. Review of the analysis and the measures to be taken 195
    9.4. Model-based analysis 195
    9.5. Limitations of the FMEA 197
        9.5.1. Common cause failures 197
        9.5.2. Other difficulties 197
    9.6. Examples 198
        9.6.1. Desk lamp 198
        9.6.2. Chemical process 199

Chapter 10. Deviation Analysis Using the HAZOP Method 201

    10.1. Introduction 201
    10.2. Implementation of the HAZOP method 201
        10.2.1. Preparing the study 202
        10.2.2. Analysis of the study nodes 203
        10.2.3. Causes and consequences of the deviation 206
        10.2.4. Result tables 208

10.3. Limits and connections with other methods 208
10.4. Model-based analysis 209
10.5. Application example 210

Chapter 11. The Systemic and Organized Risk Analysis Method . . 211

11.1. Introduction 211
11.2. Implementation of part A 214
    11.2.1. Modeling of the installation 214
    11.2.2. Identification of the hazard sources 215
    11.2.3. Building the scenarios 220
    11.2.4. Assessment of the severity of the scenarios 222
    11.2.5. Negotiation of the objectives 222
    11.2.6. Proposing the barriers 223
11.3. Implementing part B 224
    11.3.1. Identifying the possible dysfunction 225
    11.3.2. Building the fault tree 225
    11.3.3. Negotiation of quantified objectives 226
    11.3.4. Barrier quantification 226
11.4. Conclusion 228

Chapter 12. Fault Tree Analysis 229

12.1. Introduction 229
12.2. Method description 230
12.3. Useful notions 231
    12.3.1. Definitions 231
    12.3.2. Graphical representation of events and connections 232
12.4. Implementation of the method 234
12.5. Qualitative and quantitative analysis 237
    12.5.1. MOCUS algorithm 238
    12.5.2. Probability calculations 239
    12.5.3. Importance measures 241
12.6. Connection with the reliability diagram 242
12.7. Model-based approach 243
12.8. Examples 244
    12.8.1. Desk lamp 244
    12.8.2. Chemical process 244
12.9. Common cause failure analysis 247
    12.9.1. Introduction 247

12.9.2. Identification of common causes 248
12.9.3. Common cause analysis 249
12.9.4. The β-factor method 250

Chapter 13. Event Tree and Bow-Tie Diagram Analysis 253

13.1. Event tree 253
  13.1.1. Description 253
  13.1.2. Building the event tree 254
  13.1.3. Conversion into a fault tree 257
  13.1.4. Probability assessment 258
13.2. Bow-tie diagram 259
  13.2.1. Description 259
  13.2.2. Assessment of the probability 261
  13.2.3. Conversion into a fault tree 262

Chapter 14. Human Reliability Analysis 263

14.1. Introduction 263
  14.1.1. Objectives and context 263
  14.1.2. Definitions 265
14.2. The stages of a probabilistic analysis of human reliability . . . 267
14.3. Human error classification 269
  14.3.1. Rasmussen's Skill – Rule – Knowledge (SRK)
          classification 270
  14.3.2. The Reason classification 271
  14.3.3. Errors of omission and commission 272
  14.3.4. Pre-accidental and post-accidental errors 273
  14.3.5. Classification based on a cognitive model of the activity . 274
14.4. Analysis and quantification of human errors 274
  14.4.1. Performance influencing factors 274
  14.4.2. Error probability assessment 277
14.5. The SHERPA method 278
14.6. The HEART method 280
14.7. The THERP method 282
14.8. The CREAM method 288
14.9. Assessing these methods 291

Chapter 15. Barrier Analysis and Layer of Protection Analysis . . 293

15.1. Choice of barriers 293
15.2. Barrier classification 295

15.3. Barrier analysis based on energy flows 297
15.4. Barrier assessment 299
15.5. Safety instrumented systems 301
   15.5.1. Introduction 301
   15.5.2. IEC 61508 standard 303
   15.5.3. Failures of an SIS 304
15.6. The LOPA method 307
   15.6.1. Description 307
   15.6.2. Scenario identification 311
   15.6.3. Analysis of the scenarios 313
   15.6.4. Identification of the frequency of initiating events 313
   15.6.5. Identification of the safety barriers 315
   15.6.6. Calculating the risk level of a scenario 316
   15.6.7. Example 317
   15.6.8. Conclusion 318

PART 4. APPENDICES 319

Appendix 1. Occupational Hazard Checklists 321

Appendix 2. Causal Tree Analysis 327

Appendix 3. A Few Reminders on the Theory of Probability 329

Appendix 4. Useful Notions in Reliability Theory 335

Appendix 5. Data Sources for Reliability 341

Appendix 6. A Few Approaches for System Modelling 347

Appendix 7. Case Study: Chemical Process 355

Appendix 8. XRisk Software 361

Bibliography 363

Index 369

Foreword

From time immemorial, man has been confronted with danger. This danger may arise from external events (floods, earthquakes, volcanic eruptions, diseases, etc.), activities of an individual (hunting, working, leisure activities, etc.), activities of others (e.g. traffic accidents) and, more generally, from the technology mankind has developed over time (electricity, dams, cars, chemical factories, etc.).

Therefore, mankind has learned to limit these dangers, manage the associated risks and, consciously or otherwise, accept the fact that we must live with the residual risks.

Over the centuries, the part played by technology and the extension of the spatial and organizational dimensions of potential accidents and connected decisions (as shown by the recent Fukushima disaster) has led society (both on interpersonal and organizational levels) to implement risk management procedures, first to protect human lives and goods (including means of production) and second to preserve the environment.

These processes include the establishment of suitable organizational structures, the application of knowledge acquired through experience, the detection and interpretation of weak signals, training and/or raising awareness among groups at risk, the management of modifications to objects or situations that present risks, the surveillance of evolving risks (notably due to aging processes in dangerous equipment), crisis management and the control of the process itself.

Risk analysis is a key element of the risk management process. The aim of risk analysis is to identify potential accidents, examine their potential consequences and estimate the probability of occurrence. Using formal or informal acceptability criteria, a decision is taken regarding the steps required for these risks to become acceptable, potentially involving technical, human and/or organizational measures. The aim of the process is to ensure that the system in question corresponds adequately to fixed security objectives from technical, human and organizational standpoints.

The aim of this book is to provide a didactic overview of the risk analysis methodologies used in a variety of industrial sectors, with a particular focus on the consideration of human aspects. It also provides the definition of basic notions and principles associated with risks and risk management, while clearly placing the discipline of risk analysis into the broader context of risk management processes.

Sylvain CHAUMETTE

Director of the Integrated Management and
Analysis Pole of the Accidental Risks Department
Ineris
June 2013

PART 1

General Concepts and Principles

Chapter 1

Introduction

1.1. What is risk management?

What do we mean when we speak of *risk*? Let us consider the following dictionary definitions:

> – *Shorter Oxford English Dictionary*: 1. Hazard, danger; exposure to mischance or peril. 2. The chance or hazard of commercial loss, specially in the case of insured property or goods.
>
> – *Merriam Webster Dictionary*: 1. Possibility of loss or injury: see PERIL. 2. Someone or something that creates or suggests a hazard.
>
> – *Chambers Dictionary*: 1. The chance or possibility of suffering loss, injury, damage, etc; danger. 2. Someone or something likely to cause loss, injury, damage, etc. 3. (insurance) a. the chance of some loss, damage, etc., for which insurance could be claimed; b. the type, usually specified, of such loss, damage, etc., fire risk; c. someone or something thought of as likely (a bad risk) or unlikely (a good risk) to suffer loss, injury, damage, etc.

Using these definitions, we see that the word "risk" may denote a *situation of exposure to hazard*, from which *damage* may result. The notion of risk is thus connected to the notion of *hazard*, a hazard being that which may produce *damage* in the future, in an *uncertain* manner. This definition will be considered in more detail and in a more formal manner in Chapter 2.

This notion of risk is closely linked to human activity, and to human existence in general. Humanity has always been exposed to risks and humans have always generated risks to their environment; efforts to manage these risks came as a natural consequence. These risks have evolved over time, and the attitude taken to risk has evolved in parallel.

In the world of industry, risks need to be mastered for ethical, regulatory and economic reasons. This is the purpose of risk management, which, within a framework specific to each company, consists of:

– identifying risks;

– analyzing risks, that is, studying their consequences and the possibility of their occurrence;

– evaluating and ranking these risks;

– defining a strategy to use with each risk: acceptation or toleration, elimination, reduction, transfer or sharing between multiple actors.

This process is sometimes complex and is often carried out in an iterative manner. The risk management process must also make optimal use of company resources.

The aim of this book is to present the methods habitually used to implement risk management in the context of the production of goods or services. As this type of activity can generate a considerable number of more or less interconnected risks, we will concentrate on certain specified risks.

1.2. Nature of risks

Within the context of a business, we may be faced with a wide variety of risks [DAR 12]. These risks can be grouped into two categories, based on whether they only generate loss or both loss and gain at the same time:

– pure risks only present possibilities of *loss*. They are a result of undesirable events. Their occurrence creates losses for the business, while their non-occurrence does not constitute a gain, and the cost of the damage they can entail will not, *a priori*, increase. Risks associated with the security of goods and human life fall into this category;

– speculative and controlled risks can generate *losses* or *profits*, depending on events and decisions. One example of this type of risk can be found in the management of a company or a project. Decisions need to be taken involving risks. The goal is to increase profit, but a possibility of loss exists. These risks are accepted as they are the result of a choice.

The risks encountered in a business context may also be classified according to the nature of their consequences. For example, we may identify:

– risks with consequences for human health, physical or mental, generally concerning company employees, but also those living in the vicinity of sites of production;

– risks to the social and economic situation of personnel;

– environmental risks that create undesirable effects on the natural environment;

– risks to the mechanisms of production caused by phenomena within or external to the business, including natural phenomena such as flooding or earthquakes;

– risks that may damage commercial relationships, caused, for example, by malfunctions in the production mechanisms, in terms of quality, quantity or time delay;

– judicial risks that may undermine the moral entity constituted by the company, which may be held responsible for damages and thus be the target of judicial proceedings. Based on the nature of the case, we can distinguish between affairs of civil responsibility, in which another entity is subject to damage, intentional or otherwise, and criminal cases, linked to regulatory infractions. The person held responsible in these cases may be the company director, other members of the company or the company itself as a distinct moral entity. In the context of criminal cases, responsibility cannot be transferred using insurance;

– financial risks, with a direct negative impact on company assets.

Note that most of these risks have indirect financial consequences. This is the case, for example, when company goods are destroyed or damaged (in the case of major risks), or in situations where the quality or quantity of production is affected. This also applies to data security, problems of continuity in activities, problems connected with intrusion, etc. Risks of a judicial nature can lead to fines that must be paid, and risks to human health or the environment can result in the payment of damages, although these risks cannot simply be reduced to their financial aspect.

In this book, we will concentrate on risks linked to the mechanisms of production, that is those which create damage as a result of undesirable behaviors in the mechanism. The direct consequences of this type of risk concern human health, the environment and the quality and quantity of production throughput.

NOTE 1.1.– This risk is generally, although not solely, a pure risk. Take, for example, the case of a business using a manufacturing process that presents risks due to the nature of one of the products being used, for example a toxic product that could cause intoxication in humans if not sufficiently contained. A company might wish to adopt an innovative procedure to increase production. The risk linked to the danger inherent in the procedure is a pure risk. The risk linked to the decision to choose the new procedure, however, is a speculative risk, and the risk connected with the use of the site is a controlled risk.

1.3. Evolution of risk management

The methods presented in this book were developed from the 1950s onward in order to respond to a demand for greater mastery of risks, whether at company or society level. To replace these methods in their context, we will now provide a brief overview of the development of approaches to risk management.

The word "risk" has its origins in the Greek substantive "$ριζα$", meaning "root", which gave us the Latin "resecare", meaning "to cut". This, in turn, evolved to produce "resecum" in medieval Latin, meaning "reef", in a maritime context. This led to the following interpretation: the reef is an obstacle that the navigator must, imperatively, avoid.

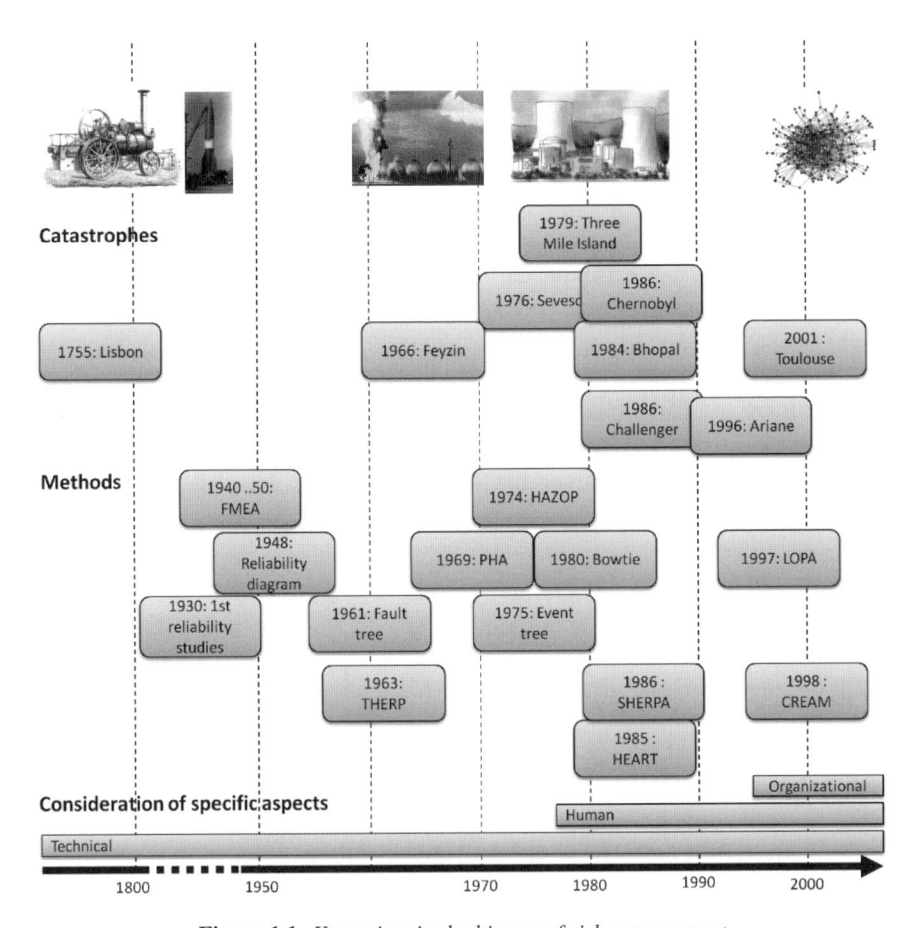

Figure 1.1. *Key points in the history of risk management*

The Lisbon earthquake of 1755, which was followed by a fire and a tsunami, constituted a key event in the development of risk management. A considerable part of the city was destroyed, and between 50,000 and 10,0000 people lost their lives. Faced with this catastrophe, Voltaire placed the blame on nature; Rousseau, however, retorted that "it was not nature which built twenty thousand six- or seven-story houses in that location". Rousseau considered that the problem was due to an error in urban development, implying that risk was not simply the responsibility of the gods, but also that of man.

Later in the 18th Century, Bernoulli, working on the probability theory initiated by Pascal and Fermat in 1654, established the law of large numbers and formulated his decision theory, introducing the notion of costs weighted by probability.

The 19th Century was marked by the Industrial Revolution, which generated industrial accidents with strong impacts on persons and on the environment. Rail transport also posed security problems, with a first important set of security regulations being established in 1893 (the Railroad Safety Appliance Act).

It was not, however, until the 1930s that the first reliability studies were carried out on the life expectancy of rolling bearings for railroads [VIL 97]. The Weibull distribution appeared in 1939 [WEI 39]. Approaches based on reliability were developed over the course of World War II, during which Lusser and Von Braun's works on the reliability of the V1 and V2 rockets were used to establish a law of reliability for a set of components in series, casting doubts on the weak link law proposed by Pierce in 1926. The notion of failure rate also emerged at this time.

The failure mode and effects analysis (FMEA) method appeared in the late 1940s, first in the military and aeronautical fields.

During the 1950s, with the growing complexity of electronic systems, the Advisory Group on Reliability of Electronic Equipment (AGREE) recommended that reliability should be integrated into the development process in order to promote the design of more reliable equipment. The advisory group also recommended the calculation of indicators such as mean time to failure (MTTF), mean time to repair (MTTR) and mean time between failures (MTBF).

Toward the end of the 1950s, a number of projects demonstrated the importance of human error in system failures. The first analytical forecasts of system reliability including human error and its quantification were published from 1958 onward [WIL 58]: these studies considered human operators as a technical component.

In 1961, H. A. Watson, working at Bell laboratories, developed the fault tree method, allowing the description of the part played by chance or hazard in the operation of complex systems.

The 1970s and 1980s were marked by a number of significant industrial and technological catastrophes:

– Flixborough, 1974: explosion of 50 tons of cyclohexane in a factory producing caprolactame, an intermediary product used in producing nylon, claiming 28 victims.

– Seveso, 1976: a cloud containing dioxine escaped from a reactor in the ICMESA chemical factory, in the town of Meda, and spread across the Lombardy plain in Italy. Four settlements, including Seveso, were affected, with significant consequences on the environment and public health. This event raised public awareness in Europe and resulted in the publication of the SEVESO directive.

– Three Mile Island (TMI), 1979: fusion of a nuclear reactor in the nuclear power plant at TMI, Pennsylvania (United States), with the release of a significant quantity of radioactivity into the environment. The incident was widely reported at international level, and had a major impact on public opinion, particularly in the United States.

– Bhopal, 1984: 40 tons of toxic gas leaked from the Carbide Union pesticide factory in Bhopal (India), killing 8,000 in the first 3 days alone. In total, the leak was responsible for more than 20,000 deaths over a period of almost 20 years. The Bhopal disaster is the most extreme example of a chemical industrial catastrophe to date.

– Challenger, 1986: a solid rocket booster exploded on take-off as a result of a leak caused by a defective O-ring seal. The crew was killed in the explosion. The technical problem was caused by design and organization failures.

– Piper Alpha, 1988: explosion of a North Sea oil rig following a gas leak, causing more than 150 deaths. The accident analysis revealed communication problems during maintenance procedures.

To respond to these major security issues, the industrial world turned to methods developed for electronic systems and in the aeronautic and aerospace domains to study the risks involved in their production facilities. In 1975, the Wash400 report, concerning safety studies in a nuclear power station [NUR 75], introduced the concept of event trees. The report also included fault tree modeling, the use of expert opinions, the inclusion of human error and feedback analysis. One of the highlighted scenarios corresponded to the TMI catastrophe. The first application of the report to an industrial site was in the context of a study of the Canvey Island complex in 1978 [HSE 78].

From the beginning of the 1980s, operational security techniques were extended to the software domain, where questions of reliability were becoming important as the field underwent rapid expansion. In the context of software design, a number of techniques were developed to enable rapid development of software with maximum reliability. Analysis, design and development procedures focusing on programming languages and methods were defined progressively, in conjunction with formal methods used to guarantee correct operation of software.

During the same period, new methods were developed to analyze the reliability and availability of systems, taking account of dynamic aspects using Markov chains and Petri networks.

The 1980s also witnessed the development of several new methods for including the human factor, such as the technique for the human error rate prediction (THERP) method (see Chapter 14).

In the course of the 1990s, risk analysis developed in a number of domains, including the automotive industry, civil engineering and building. Work was also carried out on the impact of organization, with the addition of an organizational aspect to considerations of human factors. The theories of normal accidents [PER 99] and high reliability organizations [ROB 90] were also developed at this time.

In France, the beginning of the 21st Century was marked by the AZF catastrophe, where a stock of ammonium nitrate exploded at a factory in Toulouse. This led to the creation of the law of 30th July 2003 on the prevention of technological risks. In parallel, the development of safety instrumented system and their generalization within the framework of the IEC 61508 standard led to the use of new methods, such as the bow-tie diagram at the center of the Accidental Risk Assessment Methodology for Industries in the framework of the Seveso II Directive (ARAMIS) method or the layer of protection analysis (LOPA) method, which allows a probabilistic study of accident scenarios and an evaluation of the effectiveness of security barriers.

In parallel to the development of these methods, a strategy for reflection was established using a subjective, rather than objective, vision of risk. Using an objective vision, the level of risk was considered to be independent of the observer, that is as a value that may be measured in a unique and universal manner by any observer with the requisite knowledge. This approach was re-examined by certain authors [REN 92, SLO 01, DEN 98], who proposed a subjective vision in which risk ceases to be an objective element, but rather a perceived element: the level of perceived risk depends on the observer, on what they consider to be reliable or otherwise, their intuition, their culture, media influence, etc. This is true of the general public, but is also applicable to experts. Moreover, despite the apparent objectivity of mathematical and probabilistic approaches, results are rarely used in their raw form in decision making, as a significant degree of uncertainty exists concerning available data [REI 99]. It is a mixture of the two approaches that can be found in Ren [REN 98].

In the context of this book, we will retain this latter point of view. The methods we will present are those generally used for risk analysis in structures producing goods and/or services. They allow us to obtain a measurement of risk, with a representation of the level of risk as a position on a probability–severity diagram (Figure 1.2). The results obtained in this way are not absolute, and should be interpreted with care. However, these results constitute an element for risk analysis which may serve as a point of reference, notably from a regulatory standpoint.

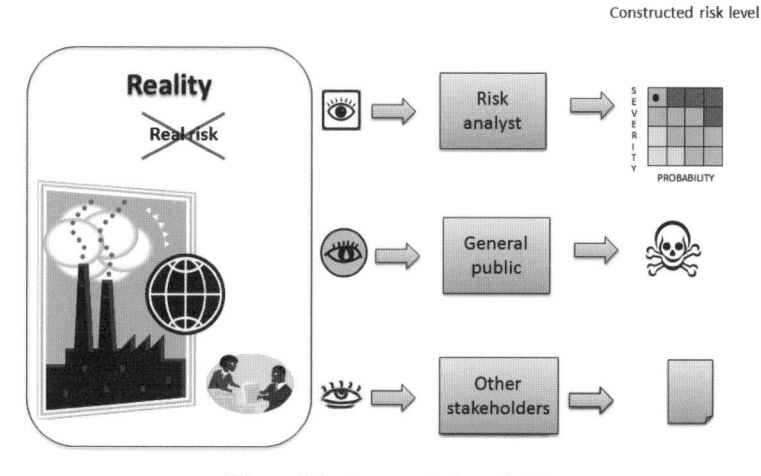

Figure 1.2. *Representation of risk*

1.4. Aims of this book

The objective of this book is to present the methods used for risk analysis in production systems. We will begin by presenting a certain number of basic notions, and then the general principle of risk analysis. Following on from this, we will examine the ISO31000 standard, which provides a specification for the implementation of a risk management approach.

The ability to represent the information we use is crucial, so we will also consider the representation of knowledge, covering both information concerning the risk occurrence mechanism and details of the system under scrutiny.

We will then present different analysis methods, first for the identification of risks, then for their analysis in terms of cause and effect and finally for the implementation of security measures.

Chapter 2

Basic Notions

In this chapter, we will present the definitions and basic principles used in risk analysis. We will consider the implementation of these different concepts in greater detail in the following chapters.

2.1. Formalization of the notion of risk

The definition of the term *risk* in everyday language requires greater precision. A variety of definitions have been proposed, including:

1) Risk is defined by a "combination of the probability of an event and its consequences" or a "combination of the probability of damage and its severity", (*definition taken from the glossary of technological risks* [MEE 10] based on the ISO51 standard [ISO 99]).

2) Risk is the uncertain consequence of an event or an action on something with a given value. It always refers to a combination of two elements: the probability or chance of potential consequences and the severity of these consequences, linked to human activities, natural events or a combination of the two. Consequences of this type may be either positive or negative, depending on the values associated with them (*definition used by the International Risk Governance Council* [GRA 05]).

3) Risk is the effect of uncertainty on objectives (*standard ISO31000:2009, [ISO 09]*):

– Note 1: an effect is a positive and/or a negative deviation in relation to an expectation.

– Note 2: objectives may have different aspects (including, e.g., financial goals, health and security aims or environmental aspects) and may operate on different levels (strategic level, project level, concerning a product, a process or an entire company).

– Note 3: a risk is often characterized in relation to events and potential consequences, or a combination of the two.

4) Risk is the combination of the probability and the significance of damages [ALE 02].

5) Risk is the two-dimensional combination of events/consequences and the associated uncertainties [AVE 07].

6) Risk is defined by a set of triplets $< S_i, P_i, C_i >$ where S_i is the ith scenario, P_i is the likelihood of the scenario and C_i is the consequence of the scenario [KAP 81].

7) A risk is a measurable danger to precise goods or activities, with harmful economic consequences [CHA 06].

From these definitions, with the exception of definition 7, we see that risk is defined in relation to *consequences*, characterized by a positive or negative *cost*, and the appearance of these consequences is dependent on an *event*, the occurrence of which is more or less *realistic*, or more or less *probable*. The two values, *importance of consequences* and *likelihood of their occurrence,* may be used to characterize the level of risk.

One of the difficulties involved in formalizing the definition of risk is the need to remain sufficiently general while maintaining a certain degree of precision. The definition of the ISO31000 standard is intended for application to all of the many and varied types of risk encountered in a business context [DAR 12]. The definition used in the standard is rather abstract and can prove difficult to interpret. It can even be difficult to understand in its definition of risk as *the effect of* uncertainty. The accompanying notes help to anchor the definition in reality.

Definition 7 is used in an insurance context, and the notion of risk is assimilated to that of hazard, without taking account of the probability of occurrence. This definition is similar to the deterministic approaches used in the domain of industrial risk prior to 2003, where risks where characterized using an evaluation of the maximum consequences. This viewpoint is not suitable for optimal risk management.

For our purposes, definition 6 is the clearest and fullest of the available definitions. The three values used to define a risk present responses to the following three questions [KAP 81]:

– "what might happen?", described by the scenario S_i;

– "how likely is this?" is defined by parameter P_i;

– "with what consequences?", C_i describes these consequences.

The notion of a scenario allows us to define what might happen, while the two other, more usual, elements characterize the significance of the risk. *Analyzing a risk* may consequently be seen as the search for the three relevant values.

To keep sight of the fact that risk is not an objective data measurement [REN 98], we should add that risk may be "defined" as a *modeling* of a situation used to respond to the three questions mentioned above.

DEFINITION 2.1.– *A* risk *is defined as being the* model *of a situation with uncertain consequences using a set of triplets with the form:*

$$R = \{< S_i, P_i, C_i >\} \text{ with } i = 1, \ldots, N$$

where S_i denotes a scenario, P_i the likelihood of the scenario and C_i its consequences.

REMARK 2.1.– It is also possible to consider the hazard and the scenario leading to the appearance of the consequence-creating phenomenon without explicitly describing these consequences. The phenomenon is characterized by the intensity of the effects it may produce over a given period of time. The risk is then characterized by the pairing (probability of occurrence and intensity of effects). When assessing the importance of consequences, we must be aware of potential targets and their vulnerability.

REMARK 2.2.– In the context of industrial or occupational risks, risk is often defined as *the possibility that a hazardous phenomenon will damage a target*. This definition fits into the definition given previously, with the scenario reduced to the dangerous phenomenon → damage sequence and the consequences considered as the damage to the target in question. Moreover, the general definition allows us to specify that the characterization of a risk focuses on the severity and likelihood of damages, and *not* those of the dangerous phenomenon, although the two are linked.

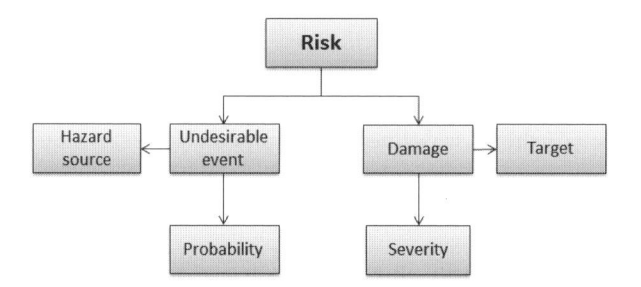

Figure 2.1. *Risk, source and target*

2.2. Hazard and sources of hazard

The notion of hazard is closely linked to the definition of risk, and it is important to distinguish between the two terms. Risk is defined as a scenario with the possibility of creating damage, whereas a hazard is the potential source generating this damage.

DEFINITION 2.2.– A hazard *is defined as the potential for damage or harm. Generally, in terms of technological risk, a hazard is associated with a system or piece of equipment involving significant energy levels, or a substance that may trigger damaging chemical or biological reactions.*

EXAMPLE 2.1.– Examples of hazards include:

– a wire carrying an electric current;

– machinery in operation;

– a virus;

– a toxic chemical product.

The *potential hazard*, the *source of hazard* or the *hazardous element* is the entity or activity comprising one or more hazards. A risk is linked to the presence of danger. As a consequence, the first stage in the risk analysis process is the identification of dangers.

In risk analysis, we need to know whether the presence of a hazard entails the emergence of a risk. Generally, we begin by defining a specific situation in which a hazard is more likely to create damage. This situation is an abnormal situation, defined either subjectively or in relation to given specifications, where the potential hazard may be released if a given event occurs. This situation is known as a *hazardous situation*. These situations involve the presence of vulnerable elements, an aspect that will be examined in more detail later.

EXAMPLE 2.2.– Some examples of hazardous situations include:

– Stripped electrical wire: generally, we consider that an electrical wire produced in accordance with modern standards is not hazardous, although it carries the potential of hazard due to the presence of an electrical current. When the wire is bare, however, damage may occur at any moment: the situation is therefore hazardous.

– The presence of a pathogenic virus in the atmosphere: this is considered to constitute a hazardous situation.

2.3. Stakes and targets

When a source of hazard releases its hazardous potential, this creates damage to elements we wish to preserve. These elements are known as stakes or targets.

DEFINITION 2.3.– *A* target *or* element at stake *is the entity affected by damage.*

Targets can be of different natures, for example:

– human targets, where damage may take the form of injury, death, chronic illness, psychological trauma, etc.;

– environmental targets, which may be destroyed or polluted;

– buildings, which may be damaged to varying degrees;

– industrial material or premises, which may be temporarily disabled, damaged or lost;

– financial assets, which can lose their value.

These targets or stakes are also known as interests to protect. They are characterized by their vulnerability and by their size or quantity.

2.4. Vulnerability and resilience

DEFINITION 2.4.– *The* vulnerability *of a target to a given effect is defined as the inability of this target to avoid damage when it is subjected to the effect in question. It may be characterized by a proportionality factor, or more generally a function between the effects to which it is exposed and the damage it incurs.*

Vulnerability depends not only on the intrinsic characteristics of the target, but also on security failings and gaps in the protection of the target.

DEFINITION 2.5.– *The* resilience *of a system may be defined as the intrinsic capacity of a system to adapt its operations before, during or following variations or disturbances in order to continue to operate, in both planned and unexpected circumstances [HOL 11].*

This definition covers the notion of robustness of a system, with the addition of an aspect concerning the adaptation of the system to new conditions. It implicitly concerns major disturbances modifying the configuration or structure of the system.

2.5. Undesirable events and scenarios

A risk is identified by seeking sources of hazard that may generate damage. The phenomenon by which damage is created as the potential hazard is released is represented using the notion of an event.

An *event* is something that may occur at a given location over a given period of time. It may also be defined as an occurrence or a change in a particular set of circumstances (ISO31000).

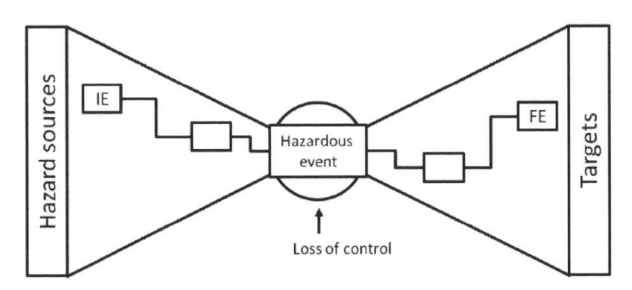

Figure 2.2. *Accident scenario represented using a bow-tie diagram*

Several types of event are used to represent an accident sequence, including hazardous events or central undesirable events (CUEs), initiating events and final events:

– A *hazardous event* (or undesirable event, or CUE, or hazardous phenomenon) is an event that describes a phenomenon capable of damaging a target. This is the first event in an event sequence, which, if left uncontrolled, will result in damage.

– A *initiating event* is an event that leads to a dangerous situation from which a hazardous event may occur.

– A *final event* is an event that represents the occurrence of damage to a target.

An *accident scenario* is a sequence of events, beginning with one or more initiating events, whose occurrence or conjunction of occurrence is necessary to produce the CUE, and which ends with the final event. This scenario may be represented using a bow-tie diagram, showing the sequence of events leading to the CUE on the left, and the series of events following the CUE on the right (Figure 2.2).

EXAMPLE 2.3.– Let us consider the description of an explosion of a sphere of gas in its liquid state. This may involve the following sequence of events:

1) Handling error (wrong valve): the initiating event.

2) Gas leak: this is the hazardous or undesirable event, as we lose control.

3) Explosion: this event is situated to the right of the CUE.

4) Destruction of goods and life: this is the final event.

We have introduced this notion of scenario into the definition of risk. The representation of a scenario as a sequence of events makes this clear.

2.6. Accidents and incidents

An *accident* can be defined as a sudden undesirable event resulting in significant damage to persons, goods or the environment.

In a similar manner, we also use the notion of an *incident*, which is defined as a sudden undesirable event resulting in minor damage to persons, goods or the environment. The distinction between incidents and accidents is partially subjective, but is usually clear in most domains of activity. In certain cases, we also speak of "near-accidents", when a sudden undesirable event takes place and damage is narrowly avoided.

This notion of a *near-accident* or *quasi-accident* appeared in a study led by Frank E. Bird Jr published in 1969 [BIR 96]. At that time, Bird was director of engineering services for the Insurance Company of North America. Bird's work showed a statistical link between the different types of accidental loss suffered by an organization in the context of their activities. For the purposes of the study, the group analyzed 1,753,498 accidents, reported by 297 companies operating in 21 different industrial sectors, representing over 3 billion manhours worked by 1,750,000 employees. The accidents were grouped into the following categories: disabling injuries, minor injuries, material damage and quasi-accidents, that is incidents with no visible damage or loss. This analysis allowed Bird to establish a ratio between different categories, represented in the form of a triangle, known as *Bird's triangle* (Figure 2.3). One of the key points of this study was that it showed that a risk management approach must also take account of quasi-accidents or even latent causes.

NOTE 2.1.– Note that this definition of an accident is not suitable for analyzing chronic risks, which cause illness as a result of prolonged exposure. In such cases, the event cannot be seen as "sudden". The damage caused presents itself in the form of a disease.

2.7. Safety

Safety is defined in everyday language as the absence of hazards capable of causing damage. This definition involves an element of subjectivity. A more precise definition is given in the standard MIL-STD882E, where the notion of safety is defined as characterizing a situation in which there is no possibility of harm to human health, damage to or loss of goods, or negative environmental impacts.

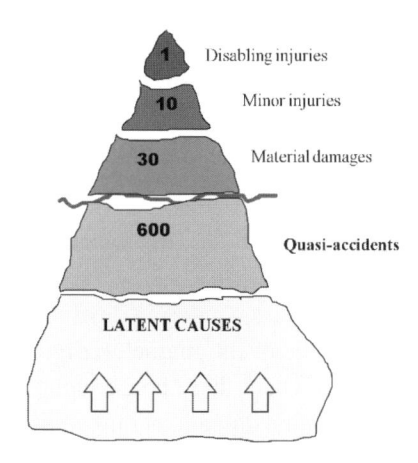

Figure 2.3. *Bird's triangle*

Using this definition, safety implies that no hazard can exist in a given situation, meaning that no risk of damage is present. This is the sole means of guaranteeing that no danger will lead to damage, and represents a major constraint. For this reason, we generally prefer to define *safety as the characteristic of a situation or a system for which there is, at a given instant, a risk of damage, which is considered to be acceptable*, or a level of risk, which is as low as reasonably achievable (ALARA), a notion that is defined in section 2.17.

2.8. Likelihood, probability and frequency

Likelihood is defined by the ISO31000 standard as the possibility that something will happen. This "something" is often represented by an event. The likelihood is thus the possibility of occurrence of this event.

This likelihood may be evaluated in a *qualitative* manner, for example:

– V1: highly unlikely;
– V2: unlikely;

– V3: possible;

– V4: likely, almost certain.

It may also be evaluated in a *quantitative* manner, generally using a number between 0 and 1. The meaning of the terms used to define levels must be precisely defined in order to avoid subjective interpretations. Where possible this likelihood is chosen based on the *frequency* of occurrence of the scenario, or its *probability*. More details on this topic may be found in Chapter 6.

2.9. Severity and intensity

The occurrence of an undesirable event characterizing a risk has a certain number of consequences. In cases where these consequences are negative, the value used to measure the scale of this damage is known as the *severity* of the risk, and constitutes a measurement of the cost of losses. It is linked to the *intensity* of the damage-creating phenomenon.

These consequences may be grouped into several categories, such as consequences to human health, consequences for the natural environment or consequences to goods. The severity can be evaluated using qualitative levels, for example:

– G1: catastrophic;

– G2: critical;

– G3: major;

– G4: minor.

Severity may also be represented by a quantitative value, representing, for example, a cost or a number of victims. It may even be described using a vector of qualitative or quantitative values if it concerns different aspects, such as damage to human health, the environment or economic losses; this is the case for the European accident scale (Figure 6.6). As in the case of likelihood, it is important to provide precise definitions of terms in order to avoid subjective interpretation. Chapter 6 provides a detailed presentation of this notion and of different scales.

2.10. Criticality

The level of a risk is obtained by combining its likelihood and the severity of the consequences. This combination may be carried out in a quantitative manner by introducing the notion of *criticality*:

$$c = p.g$$

where p is the likelihood and g characterizes the severity. If p represents a probability, c may be seen as the mathematical expectation of loss measured by the severity. This value c allows us to propose an approach to define a notion of tolerable risk by fixing a maximum criticality threshold (Figure 2.4), known as the Farmer criterion.

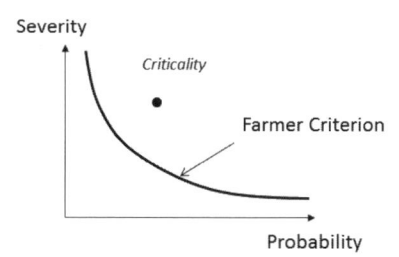

Figure 2.4. *Farmer criterion*

The choice of this tolerable threshold is a key element in risk assessment. This choice may be imposed by regulations, as in the case of major industrial risks [MEE 10]. In other cases, a choice is made by consensus between interested parties, or by using an approach such as the as low as reasonably practicable (ALARP) principle, discussed in section 2.17.

When a risk is evaluated using qualitative values, it is characterized by a pairing *(p,g)*. To define a risk level, we use a *risk matrix* (Figure 2.5). Different levels of risk are defined, such as very high, high, medium and low. The matrix presented in this chapter provides an illustration of this notion. Details for effective implementation will be given in Chapter 6.

2.11. Reducing risk: prevention, protection and barriers

Risk reduction techniques are used to modify the level of a risk. These techniques take varying forms, including the use of technical equipment, instructions given to personnel, training or measures taken at an organizational level. The risk that remains after these techniques have been implemented is known as the *residual risk*.

	V1 Extremely unlikely	V2 Unlikely	V3 Possible	V4 Likely
G4 Catastrophic	Medium	High	Very high	Very high
G3 Critical	Medium	High	High	Very high
G2 Major	Low	Medium	High	High
G1 Minor	Low	Low	Medium	High

Figure 2.5. *Risk matrix*

Two broad types of action may be taken to modify the level of a risk:

– We may seek to reduce the likelihood of occurrence of a hazardous event, that is to reduce the possibility of occurrence of the risk. In this case, we speak of preventive measures.

– Alternatively (or additionally), we may seek to reduce the scale of damage so as to reduce the severity of the risk if a hazardous event occurs. In this case, we speak of protective measures.

EXAMPLE 2.4.–

1) Let us take the example of a workshop with containers of highly toxic products. The first approach might include regular checks on the state of the containers to avoid leakage. This is a preventive measure. Another approach would involve sealing the workshop so that, if a leak did occur, the severity of the consequences for the vicinity would be reduced. This is a protective measure.

2) Let us consider a cyclist, at risk of serious injury in the case of a fall. A preventive measure would be to train the user to take care when cycling. Asking the cyclist to wear a helmet would be a protective measure.

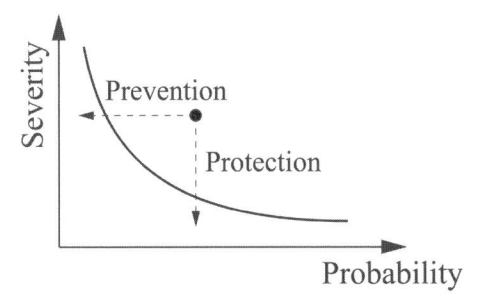

Figure 2.6. *Reduction of probability and severity*

The difference between preventive and protective measures can be seen in relation to the CUE: either we prevent the event from occurring, or we prevent the final consequences from appearing.

In certain situations, risk reduction techniques may be seen as being either preventive or protective depending on the chosen viewpoint. For this reason, the term "safety barrier" is often used and may be positioned on a bow-tie event diagram (Figure 2.2). This representation allows us to clearly show events with a reduced likelihood, and the residual events, of lesser importance, which may still occur when the barrier is active.

These aspects will be covered in greater detail in Chapter 15.

2.12. Risk analysis and risk management

The standards published in recent times, notably the ISO31000 standard and the ISO 73 guide, propose a definition of risk management as a process including:

1) a risk identification aspect: this process involves finding, recognizing and describing risks. Each risk is described by identifying sources of hazard, events that would release the potential hazard and consequences;

2) a risk analysis stage, which allows us to understand the risk and evaluate its level;

3) an evaluation aspect, where we determine whether or not a risk is acceptable with regard to defined criteria;

4) a final risk treatment stage, which consists of modifying a risk to make it acceptable: we may seek to reduce it (in terms of likelihood or severity), to transfer it (insurance) or to put an end to the activity.

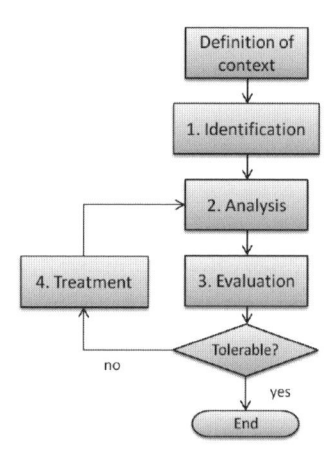

Figure 2.7. *Simplified risk management process*

The level or importance of a risk is expressed as a combination of the severity of the consequences and their likelihood, as explained earlier.

The full process also includes phases of communication, monitoring and review of implemented actions, and may be repeated in an iterative manner as part of a continuous improvement approach (Chapter 4).

Generally, we use the term "risk analysis" (stage 2 in the process described above) to describe the risk management process as a whole, and the term "risk assessment" to describe stages 1 to 3, or even the whole process when dealing with the assessment of occupational risks.

2.13. Inductive and deductive approaches

A risk analysis method may be inductive or deductive:

– In an inductive approach, we use causes to identify effects. Based on different possible faults or deviations, we seek to identify consequences for the system or its environment. Most of the methods used in risk analysis (preliminary hazard analysis (PHA), failure mode and effect analysis (FMEA), a hazard and operability study (HAZOP), event trees, etc.) are inductive.

– In a deductive approach, we identify effects and work backward to find their causes. The system is presumed to be in a state of failure, and the analytical process seeks to identify the possible causes leading to this state. The fault tree method of analysis is a deductive method. The use of causality trees when analyzing accidents also constitutes a deductive approach.

2.14. Known risks and emerging risks

A *known* risk corresponds to a known hazard with known damaging effects. A fire risk, for example, is a known risk.

Emerging risks concern situations that may generate new risks (new technologies, new contexts, newly exposed populations, etc.) and where the possible damage, notably effects on human health, is not always well known or understood. Preventive measures in relation to these new risks are based on alert processes, research and development, and raising public awareness. The potential risks associated with nanoparticles or with mobile telephony fall into this category.

REMARK 2.3.– Article 34 of the European ruling no 178/2002 adds to this definition of emerging risks, including risks that are already known but where exposure is increasing, and known risks where scientific advances have shown the severity of the consequences to be greater than previously understood.

2.15. Individual and societal risks

When measuring risks to humans, we distinguish between *individual risks* and *societal risks*:

– An *individual* risk is the probability that a person exposed to the danger will lose his/her life.

– A *societal* risk is the probability that more than a certain number of individuals will lose their lives as the result of an accident. These risks are represented using F/N curves, where F is the frequency of a given number N of victims (Figure 2.8).

If exposure to the danger is continuous, then the individual risk may be defined using the probability of death per annum:

$$T_{individual} = Pr(\text{Death of an exposed individual})$$

28   Risk Analysis

which may be estimated using

$$T^*_{individual} = \frac{\text{Number of observed deaths}}{\text{Number of people exposed}} \text{ in the course of a year}$$

Figure 2.8. *Example of an F/N curve (probability of a given number of victims)*

2.16. Acceptable risk

In the context of risk analysis, analytical methods may be used to construct a risk model, in the form of scenarios, associated with a likelihood and with consequences. The evaluation stage consists of determining acceptable risk levels based on this model.

The notion of *acceptability* is used to determine what might be considered tolerable for interested parties in the context of losses resulting from the manifestation of a risk:

– According to the OHSAS 18001 benchmark [OHS 09], an acceptable risk is a risk that has been reduced to a tolerable level for an organization in relation to its legal obligations and to its own workplace health and safety policies.

– According to ISO/IEC Guide 51 [ISO 99], an acceptable risk is a risk that is accepted in a given context, based on the current values of our society.

This notion may evolve over time, based on the level of information available to interested parties and their risk culture, and depending on countries. It constitutes a *central and sensitive issue* in the risk management process. From a technical perspective, the idea of acceptable risk is expressed through the definition of thresholds or zones on the risk matrix or the probability–severity diagram. The difficulty for the company and interested parties resides in the choice of these thresholds.

2.17. The ALARP and ALARA principles

The concept of ALARP was first introduced in the United Kingdom [HSE 01]. Using this principle, the level of risk is divided into three zones (Figure 2.9):

– Intolerable risk zone: in this zone, the risk is unacceptable, no matter what advantages may be associated with the activity; the risk must, imperatively, be reduced.

– Tolerable risk zone: in this zone, the risk is undesirable, and measures should be taken to reduce the risk, unless it is possible to show that the cost of reducing the risk is disproportionate in relation to the possible improvement.

– Globally acceptable risk zone: in this zone, the risk is accepted and no reduction measures need to be taken. The remaining available resources should be used to reduce other risks.

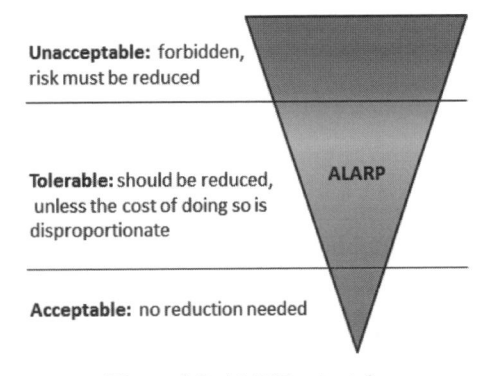

Figure 2.9. *ALARP principle*

This principle clearly shows two boundaries: an upper limit, which defines risks that must be reduced in all cases, and a lower limit, which defines risks that do not need to be reduced. The term *reasonably acceptable* takes on its full meaning in the *tolerable* zone. Using this principle, the risk level is compared to the cost involved in further reducing the risk. To avoid investing resources in reducing a risk, the entity responsible for this risk must be able to show that the cost of reduction would be clearly disproportionate in relation to the advantages obtained in terms of risk reduction. The process is not, therefore, a simple comparison between the cost and advantages of measures; instead, risk reduction measures should be adopted as a matter of course, except in cases where they entail sacrifices which are clearly disproportionate. An extreme example [HSE 01] would be:

– spending one million pounds to improve the working conditions of five employees experiencing knee pain would clearly be disproportionate;

– spending one million pounds to avoid a major explosion capable of killing 150 people, however, is clearly entirely appropriate.

Thresholds have been established in the United Kingdom in relation to the annual probability of death for an individual:

– the maximum threshold, between tolerable and unacceptable zones, is set at 10^{-3} for employees, 10^{-4} for the population in the vicinity for an existing site and 10^{-5} for this population in the case of a new site;

– the lower threshold, between acceptable and tolerable zones, is set at 10^{-6} for the population in the vicinity of a site.

Other similar principles have been developed, including *ALARA* in the Netherlands or GAMAB (*Globablement au moins aussi bon*) in France [SCH 01]. The ALARA principle is conceptually close to the ALARP principle. However, the maximum acceptable risk threshold is set at different values:

– 10^{-6} per year for the public in the vicinity of an industrial site, instead of 10^{-4};

– there is no lower threshold in the ALARA principle: all risks must be reduced, as long as such measures remain reasonable.

The GAMAB principle is much less widespread. According to this principle, the total risk associated with a new system must be, at most, equal to that of existing systems.

The French approach is, in fact, close to the ALARA principle. An acceptable risk is that which is legally permitted (below the upper boundary), where benefits in terms of risk reduction continue to outweigh the costs of implementing reduction measures.

2.18. Risk maps

A risk map is a diagram representing all of the risks present in a system, an installation or an organization. It often takes the form of a probability–severity diagram or a risk matrix (Figure 2.10) showing the main risks. It is sometimes useful to have more synthetic diagrams, of a radar or other type (Figure 2.11).

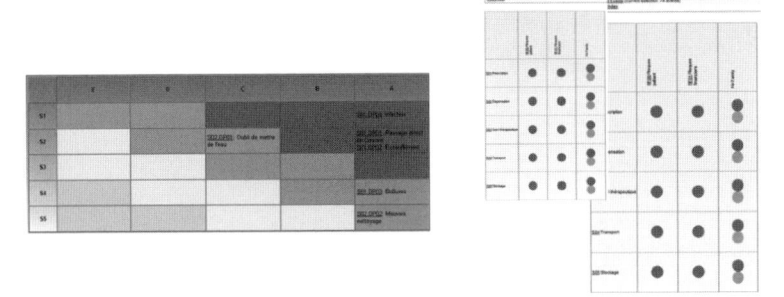

Figure 2.10. *Risk map*

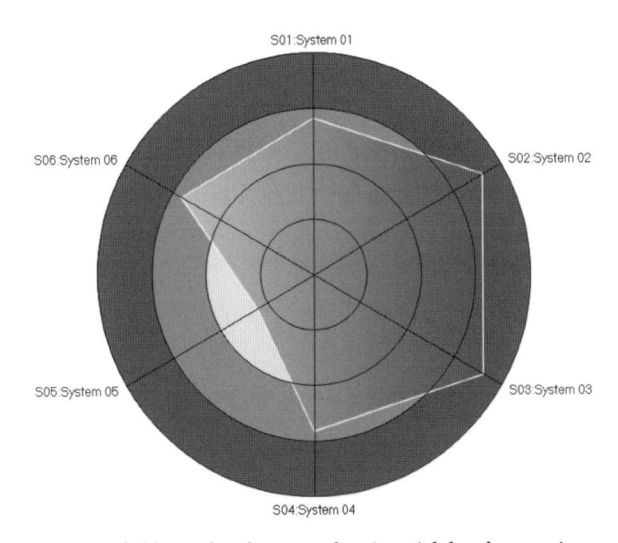

Figure 2.11. *Radar diagram showing risk level per unit*

Chapter 3

Principles of Risk Analysis Methods

3.1. Introduction

As we saw in Chapter 2, the notion of risk, as used in this book, refers to an event that may occur in the future, with damaging consequences for the system in question. In this context, the general approach used in risk analysis is as follows:

1) What damage is possible, and to what targets?

2) What are the sources of hazard and undesirable events, which may generate this damage?

3) In what situation(s), considered to be hazardous, might this event occur?

4) What are the causes of existence or appearance of these hazardous situations?

5) Do barriers exist to prevent an accident if the undesirable event does occur?

Steps 1 and 2 allow us to identify hazards, whereas steps 3, 4 and 5 allow us to evaluate the level of risk:

– Step 1 consists of defining potential targets, which may be implicit, and the type of damage under consideration.

– Step 2 consists of identifying hazard sources, that is elements or activities that may *create* damage, i.e. that are the direct root of the damage-causing mechanism if an undesirable event occurs.

– A target under threat from a hazard source that may, at any moment, create damage if a certain event occurs constitutes a *hazardous situation*. The *severity* of the situation may be evaluated based on potential damage, taking account of the vulnerability of targets.

– To evaluate the *likelihood* of a hazardous situation, we analyze the causes that make this situation possible. These causes may be linked to technical or human aspects, and may themselves be the result of organizational aspects or external hazards.

– The final stage consists of analyzing existing barriers, both at preventive and protective level in order to determine their *efficiency* and the level of effective risk.

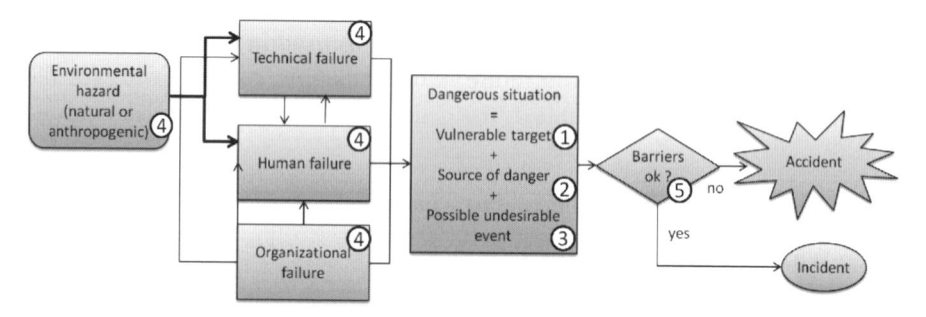

Figure 3.1. *General diagram of the analysis process*

This approach as a whole is based on:

– general knowledge of hazard sources, the associated damage and the different types of possible cause leading to the appearance of a hazardous situation; this information is organized in the form of a taxonomy;

– specific knowledge related to a domain of activity;

– methods supporting a systematic analytical approach.

In this chapter, we will consider the following aspects involved in risk analysis for production systems:

– targets and the associated damage;

– a certain widespread hazard sources;

– the causes behind the appearance of a hazardous situation, which may lead to damage.

3.2. Categories of targets and damages

A system of production of goods and/or services is a socio-technical system, made up of both a technical aspect and a human component with an organizational aspect.

EXAMPLE 3.1.– A chemical site, a unit producing household electrical goods or a hospital all constitute examples of production systems.

In these systems, the main targets are:

– People, who may be grouped into different categories:

- members of the company;

- clients, users of goods or beneficiaries of services;

- populations in the vicinity of production sites.

– Material goods:

- means all that is material in a company (building, machines, products, etc.)

- information;

- property in the neighborhood of the company.

– The natural environment.

– The entity or entities with legal and financial responsibility for the company.

Following the occurrence of an undesirable event, damage of different types may be observed:

– Concerning people, damage impacts human health, whether physiological or psychological. Damage can range from mild discomfort to death, via illness and injury. It may be caused by an accident or a somatic disease, or result from psychological causes.

– Concerning the means of production, damage has an effect on availability, from a small loss of time to the total destruction of the apparatus.

– In terms of products or services, damage can take the form of a reduction in quality, insufficient quantity or excessive time delays.

– For the manmade environment, damage can range from a disturbance in terms of normal operations to more or less complete destruction.

– Damages to the natural environment might include aquatic, atmospheric, noise or olfactory pollution.

This damage observed in and around the system of production has an impact on other components of the company and generate associated risks, notably:

– financial risks, such as the costs associated with damages to goods or to persons;

– social risks, such as lack of motivation or the stress generated by a faulty system of production;

– commercial risks, such as a loss of clients following a reduction in production capacity;

– legal risks, linked, for example, to damage to surrounding populations or the production of defective products.

Thus, it is clear that all risks to a company are interlinked, and that a global risk management approach is necessary. In the specific context of risk analysis for a production system, the first step involves defining targets under direct threat and the damage we wish to consider.

3.3. Classification of sources and undesirable events

3.3.1. *General points*

The list of hazard sources depends on the type of risk we are considering. These risks may be internal or external to the system in question. For risks generating physical damage, a widespread approach consists of seeking energy sources that are able, via a physical, chemical or biological phenomenon, to create damage. This approach, known as the *energy source* approach, is relatively general. Sources may be of several types:

– electrical (burning, electric shock, projection, electrocution, etc.);

– mechanical (crushing, slicing, catching, knocks, etc.);

– chemical (runaway reactions, untimely reactions, irritation, intoxication, disease, etc.);

– sources of ionizing (radioactive) or other radiation (light, infrared, ultraviolet, electromagnetic radiation, etc.);

– noise related (can cause deafness, tinnitus, fatigue, stress, dizziness, etc.);

– thermal (discomfort, burning, etc.);

– pneumatic (projection, etc.);

– hydraulic (projection, perforation, abrasion, etc.);

However, this approach has certain limitations:

– A description using energy sources does not always correspond to standard practice. The mechanical energy due to the weight of an object or an individual, for example, may generate a risk of falling either from a height or on the same level. In practice, terms such as "falling" tend to be used instead of "risks linked to mechanical potential energy".

– Certain phenomena are not linked to a source of energy. This is the case for biological phenomena, with pathogenic agents, and in the case of asphyxiating gases in chemistry. In the same way, risks linked to poor posture cannot clearly be linked to a source of energy.

– The description of a phenomenon using only the energy source can be too general. It is also useful to specify the undesirable event. This is the case, for example, of the source of a fire in cases of chemical risks. It is also useful to specify the conditions of activation of this source. This is also the case for risks linked to a type of activity, for example maintenance, where a variety of sources may be present.

For this reason, we often use specific lists for different types of risk and by domain of activity. We now present the lists suited for use with occupational risks and for major industrial risks. Other lists are given in Chapter 8 on the preliminary hazard analysis (PHA) method, which aims to identify undesirable events for a system.

3.3.2. *Case of occupational risks*

Occupational risks are risks that affect operators in the context of their professional occupation. The different hazard sources and undesirable phenomena can be grouped into the following categories:

– Risks linked to physicochemical phenomena and to machines:
  - mechanical risks;
  - electrical risks;
  - chemical risks;
  - risks linked to explosions and to fire;
  - biological risks.

– Risks linked to the physical environment and to working conditions:
  - risks associated with noise;
  - risks associated with air quality;
  - risks associated with lighting conditions;
  - risks associated with temperature conditions;
  - risks associated with vibrations;
  - psychosocial risks.

– Risks linked to physical activity and to the organization of work:
  - risks linked to manual handling;
  - risks linked to mechanical handling;
  - risk of falling – either tripping or from a height;
  - risks linked to working in isolation;

- risks linked to the involvement of external companies;
- road traffic risks.

Appendix 1 contains a detailed presentation of these risks.

3.3.3. *Case of major industrial risks*

A major industrial risk is an accidental event occurring at an industrial site with immediate and serious consequences for the employees, neighboring populations, goods and the environment. These risks mostly concern chemical and petrochemical compounds and the transportation of hazardous matter.

The consequences of an accident in these industries may be grouped using the following three effect typologies:

– Thermal effects, linked to the combustion of a flammable product, or to an explosion.

– Mechanical effects, linked to excess pressure, resulting in a shock wave (fire or detonation) provoked by an explosion. This may come from explosive material, a violent chemical reaction, violent combustion (combustion of gas), sudden decompression of a pressurized gas (e.g. explosion of a bottle of compressed air) or the inflammation of a cloud of combustible powder. For these consequences, specialists are able to calculate the excess pressure created by the explosion using mathematical equations. This information is used to determine the associated effects (burst ear drums, lung damage, etc.) using plotters.

– Toxic effects resulting from the inhalation of a toxic chemical substance (such as chlorine, ammoniac or phosgene) following a leak in an installation. The effects of this inhalation might include pulmonary edema or damage to the nervous system.

The Accidental Risk Assessment Methodology for Industries in the Framework of Seveso II Directive (ARAMIS) method developed by the INERIS (*Institut national de l'environnement industriel et non risqué*) [DEL 06] defines a typology of undesirable events in the context of major risks, as shown in Figure 3.2. These risks may be generated by any element of an industrial installation or in transportation.

40   Risk Analysis

No.	Description
ERC1	**Decomposition (solid substances, except bulky stocks)**: change in physical state of the substance due to the addition of energy/heat or by reaction with an incompatible chemical substance. The decomposition of the substance leads to an emission of toxic gases or to a delayed explosion of the flammable gases formed in this way.
ERC2	**Explosion (solid explosive substances in bulk)**: change in physical state of the substance due to the addition of energy/heat or by reaction with an incompatible chemical substance. The state change results in solid combustion with excess pressure (or explosion) effects resulting from a violent and spontaneous reaction. In the case of a solid stored in a closed recipient, the explosion is considered to be a cause of internal excess pressure, which may lead to a loss of confinement (catastrophic rupture or breach).
ERC3	**Release of dust by air circulation** (for example, excessive ventilation) and release of powder exposed to the atmosphere (open storage, conveyors, etc.).
ERC4	**Release of dust by liquid** (for example, flooding or overflow of a liquid) involving dust and powders exposed to the atmosphere (open storage, conveyors, etc.).
ERC5	**Inflammation – fire**: reaction between an oxidizing product and a flammable or combustible product, or a decomposition of an organic peroxide leading to fire. This ERC concerns substances where a loss of physical integrity (decomposition, contamination) leads to fire. This ERC may be associated with pyrotechnical substances.
ERC6	**Breach in gaseous phase** through a hole of a given diameter in the walls of equipment in the gaseous phase (or with a solid in suspension), leading to continuous output.
ERC7	**Breach in liquid phase** through a hole of a given diameter in the walls of equipment in the liquid phase, leading to continuous output.
ERC8	**Tube leak in liquid phase** through a hole of diameter equal to a certain percentage of the nominal diameter of a tube carrying a liquid, including "functional" openings in the tube: leaky joints at pumps, valves, etc.
ERC9	**Tube leak in gaseous phase**: this ERC corresponds to a hole of diameter equal to a certain percentage of the nominal diameter of a tube carrying a gas. The ERC can be a "functional" opening in the tube: leaky joints at pumps, valves, stoppers, etc. This ERC also applies to tubes carrying a solid in suspension in a gaseous phase.
ERC10	**Catastrophic rupture**: catastrophic rupture corresponds to the complete loss of equipment leading to instantaneous and complete output of the substance. The BLEVE is also considered as a particular catastrophic rupture. In certain cases, catastrophic rupture can lead to the ejection of missiles and a high pressure wave.
ERC11	**Collapse of container**: collapse of a container corresponds to the complete loss of equipment leading to instantaneous and complete output of the substance. The ERC is due to a reduction in pressure of the container, leading to its collapse through atmospheric pressure. This ERC does not lead to the ejection of missiles or the production of a high pressure wave.
ERC12	**Collapse of container lid**: collapse of a lid can be due to a reduction in internal pressure leading to the collapse of the moveable lid through atmospheric pressure. Specific case of air-based atmospheric storage.

Figure 3.2. *ERC table used in the ARAMIS method*

3.4. Causes of technical origin

Causes of technical origin are linked to a fault or failing in the equipment used to carry out an activity. These causes may concern material components, software, or a fluid or product used by this equipment.

The EN 60011 standard offers several classifications of failure:

– The first method is based on the type of cause. We may distinguish between the following:

- failures with intrinsic causes, also known as primary failures: failures through wear and tear (linked to the use lifespan), through aging (linked to age), failures caused by poor design, faulty production or incorrect installation of equipment;

- failures with extrinsic causes: failures linked to the incorrect use, poor handling, maintenance or consequences of another fault. This type of failure is also known as a secondary failure, or a command failure when it is a consequence of an incorrect command order.

– The second classification method is based on the degree of importance: complete, partial, permanent, fugitive or intermittent failure, etc.

– The third and final classification is based on the speed of appearance: sudden or progressive. Failures that appear progressively generally correspond to the degradation of an element that loses its properties in a progressive manner.

Extrinsic causes arise from other elements of the system, and may be of varying natures:

– *material*, for example a lack of electric current or a fault in an onboard processor;

– *human*, for example use of incorrect settings by a maintenance operator;

– *organizational*, for example a problem caused by understaffing.

The analytical process, therefore, involves looking for elements interacting with the components in question. Intrinsic causes, which we will now consider in detail, are due to the characteristics of the component itself.

3.4.1. *Material failures*

We now consider a number of possible origins of material failure.

3.4.1.1. *Mechanical failure*

Mechanical failures in systems, notably in the case of moving parts or of components subjected to constraints, may be grouped into the following four categories:

1) Mechanical failures due to surface deterioration, notably due to:

- friction, which leads to wear, that is the progressive removal of matter from the surface;

- rotation, which may lead to flaking and the loss of matter;

- abrasion, causing the loss of matter due to the action of impurities (sand, dust, loosened metallic particles, etc.).

2) Mechanical failure through deformation:

- plastic deformation due to mechanical constraints, caused by exceeding the elastic limits of the material;

- plastic deformation due to thermal constraints and over the course of time.

3) Mechanical failure through rupture:

- ductile rupture: this occurs after a phase of significant plastic deformation, stretching of the material and extrusion at the site of rupture;

- fragile rupture: this occurs after a very small plastic deformation, often the effect of shock, and is able to occur due to the intrinsic fragility of the material;

- rupture through fatigue, after a high number of uses above a threshold that depends on the material.

4) Failures involving catching or blocking for components with mobile parts.

3.4.1.2. *Failure due to corrosion*

Corrosion is the degradation of metals. Several forms of the phenomenon exist:

– electrochemical corrosion: this corresponds to an electrolysis reaction. Atmospheric corrosion falls into this category (in this case, the electrolyte comes from the water contained in the atmosphere);

– chemical corrosion, due to accidental or normal, temporary or permanent contact between equipment and aggressive products;

– electrical corrosion, an effect of leaking currents that create a difference in potential between two neighboring metallic surfaces, which is sufficient to create an electrical arc, leading to abrasion;

– bacterial corrosion: cutting oils and water for industrial use often contain "ferro-bacteria", which grow extremely quickly (one bacterium can produce 1 billion bacteria in 12 h) and corrode metals.

3.4.1.3. *Failures in electric and electromechanical components*

The main failure modes for these components are:

– rupture of electrical connections, often as a result of an external cause (shock, overheating, vibration, etc.);

– short-circuits in coils;

– sticking or wear to contact points after a certain number of cycles;

– for electromechanical components, such as bridges or actuators, failures may be linked to mechanical components and include blockages in the armature or catching.

3.4.1.4. *Failures in electronic components*

Electronic components may be grouped into passive components (resistors and capacitors) and active components (semiconductors). Failures in these components are mostly of the following types:

– Short-circuit: the component no longer presents resistance.

– Open circuit: the component prevents the passage of current.

– A change in properties of passive components, such as capacitors.

– Connection rupture with sticking at 0 or 1 for active components.

These failure modes are difficult to prevent. It is, however, possible to act on the external phenomena, which generate the failures and which are environmentally dependent. The external environment is characterized by conditions including temperature, humidity, shocks, vibration, pressure, corrosion, dust gathering and the aggressiveness of the environment in which the circuit is operating. These environmental constraints affect components and can provoke changes in characteristics, which lead to a final breakdown.

3.4.1.5. *Failures linked to electromagnetic compatibility*

Most electrical and electronic equipment generate electromagnetic fields, which may be detected in the surrounding environment. These fields create a type of pollution that can disturb the operation of other equipment. Electromagnetic compatibility (EMC) is the capacity of a piece of equipment to function correctly in an electromagnetic environment, without itself producing disturbances in this environment (IEC standard 60601):

– an equipment is said to be *disturbing* when it emits interference signals, whether through radiation (electromagnetic field) or by conduction (conducting wires and printed circuits).

– an equipment is said to be *disturbed* by radiation or conduction when a source generates interference signals leading to the failure of the equipment.

3.4.2. *Failures in software and information systems*

Nowadays, most production systems include a control system, known as a distributed control system (DCS), a supervisory control and data acquisition (SCADA) system or onboard information systems. Failures in these systems can result from material or software aspects. For the material aspect, possible failure modes are the same as those presented in the case of electronic or electromechanical components (relay circuits). Software failures may be linked to:

– Execution:

 - incorrect result produced due to software errors;

 - result supplied too late, in cases where processing capacity is insufficient for requirements.

– Data:

 - data may be lost or altered due to errors in software;

 - data may be temporarily unavailable.

Software errors are different from material errors in a number of ways:

– They are systematic: with the same starting state and the same input, the same error is produced. In the case of control systems or onboard systems, it may, however, be difficult to recreate an identical situation in order to replicate a fault.

– They do not depend on the age of software, but on its design.

To reduce design errors, software developers use quality assurance methods [SOM 10], and software may be validated using formal methods [GNE 12].

Aspects linked to viruses or hacking will not be considered in the context of industrial risk analysis, but should be taken into account in the context of information and computing security.

3.4.3. *Failures linked to fluids and products*

Faults may be caused by an absence, an excess or a change in the properties of products used in a system. An insufficient quantity of lubricant in a motor, for example, can cause a fault. Another example would be an excess of impurities in cooling fluid, which would affect the operation of the cooling system.

In a general manner, we can distinguish between:

– flows that are consumed, used or transformed by the procedure, whether fluids or flows of pieces being processed;

– utilities (liquids and energy) ensuring the normal operations of the system, supplying, for example, electrical energy, hot water, cold water, steam, compressed air or inerting gases (such as nitrogen).

For each of these flows, the main failure types are:

– an insufficient, non-existent or excessive supply;

– a change in properties.

The hazard and operability study (HAZOP) method (Chapter 10) is well suited to analyzing these changes.

3.5. Causes linked to the natural or manmade environment

The environment in which a system operates can generate hazards, which may be the direct or indirect cause of undesirable events. In the case of the natural environment, the main hazards to consider are:

– flooding;

– climatic conditions with extreme temperatures (freezing and heat waves);

– earthquakes;

– land shift;

– forest fires;

– storms;

– volcanoes, avalanches, tidal waves, etc. (in zones where these phenomena occur).

Human activities in the vicinity of a production system may also generate risks for the system in question. This manmade environment can be divided into technical and socioeconomic aspects. The main hazards to consider are:

– domino effects generated by technological accidents in a neighboring business;

– effects created by external contractors, notably in maintenance;

– more generally, effects generated by various interested parties involved in the sphere of activity of the company.

3.6. Human and organizational factors

When analyzing accidents, we often see that human and organizational factors constitute the root cause. In this context, we speak of human error, a term that we will define more precisely in Chapter 14. Human error can occur directly in the execution of an action, in the system usage phase or in other phases of the system lifecycle (maintenance, construction, etc.). It may result from a communication problem between entities, a poorly designed installation, or errors in decision making or system organization, which generate latent faults. The 2012 accident inventory [MEE 12] attributes 59%

of the causes of industrial accidents to organizational and human factors. Similar levels are often observed in other sectors.

Generally speaking, the causes of human error can be linked to person-specific individual aspects, to technological aspects or to organizational aspects [HOL 98] (Figure 3.3).

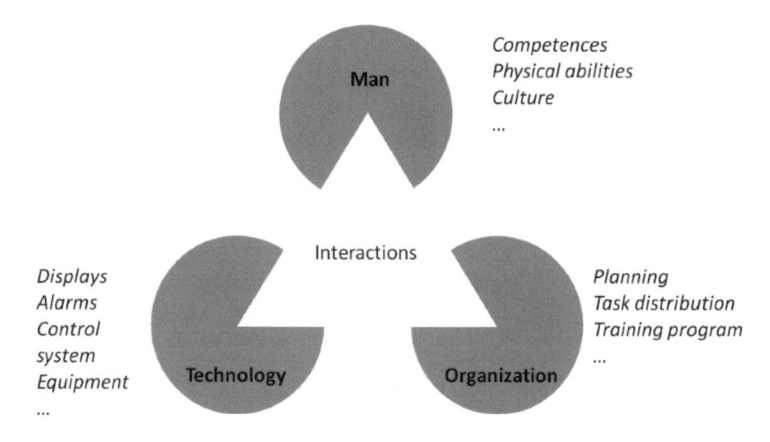

Figure 3.3. *Causes of human error [HOL 98]*

Causes specific to individuals are linked to different aspects such as their abilities, attitude, physical characteristics, psychological characteristics and culture. These aspects influence his or her behavior. A certain number of these aspects are unchangeable, but others may change and can be improved through suitable training and raising awareness.

The second type of causes of error includes those linked to interactions with material. They are involved in the reception of information (display and sounds) and in the execution of actions, whether in relation to a physical object or an action carried out using a computer system.

The third type of cause is connected to organizational aspects. These aspects have an impact on individuals and equipment, and are, consequently, often initial causes. They are intrinsically linked to company culture.

Chapter 14 presents several methods for use in studying human reliability.

3.6.1. *Reason's analysis of the human factor*

Reason [REA 97, REA 90] presented a modeling of the human factor, often known as the "Swiss cheese" modeling of human error (Figure 3.4). This modeling is often used to analyze risks not only in the industrial domain [COU 11], but also in a medical context [SHO 12]. It also served as a basis for the Human Factors Analysis and Classification System (HFACS) classification system for aviation [SHA 07]. The modeling considers four levels of causes of human error:

– *active* failures, corresponding to poorly executed human actions, labeled as hazardous acts;

– three further levels for *latent* failures:

- preconditions for unsafe acts,

- supervisory failings,

- organizational faults.

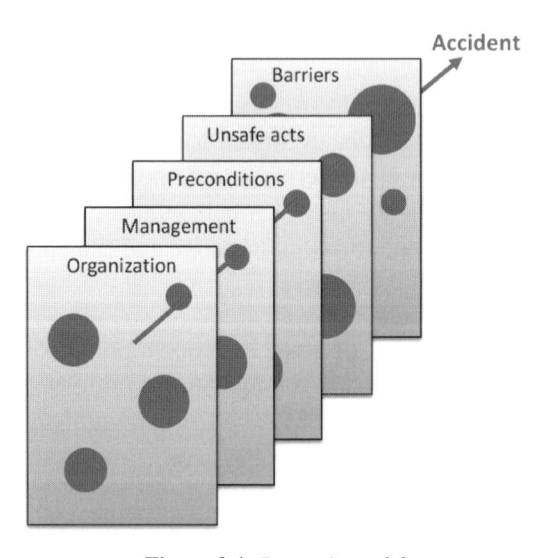

Figure 3.4. *Reason's model*

Active failures are failures that are the direct causes of an accident and correspond to error modes identified by different risk analysis methods. Latent causes often remain undetected until an accident occurs and are causes of the occurrence of a human failure.

EXAMPLE 3.2.– An example of a latent cause in an industrial process would be a set of two valves that are closed together and badly labeled. Another example, in a hospital, would be two different injectable products in similar bottles and with similar labeling. These two examples of latent causes create the possibility of an accident as a result of an active failure, in these cases a selection error.

3.6.1.1. *Unsafe acts*

Unsafe acts can be divided into two categories, errors and violations, which can themselves be split into subcategories. Errors are linked to involuntary behaviors, whereas violations correspond to deliberate infractions of rules and regulations.

We may distinguish three types of error: those linked to execution, those linked to decision making and those linked to perception. In the case of violations, we distinguish between systematic and exceptional violations.

3.6.1.2. *Preconditions*

Preconditions correspond to a state where the operator is more prone to commit errors. These include:

– aspects relating to a psychological state: stress, mental confusion, complacency, distraction, psychological fatigue, pressure to execute complex or unfamiliar tasks in a short period of time, lack of motivation, overwork (too many tasks) or the lack of attention;

– aspects relating to a physical state: fever, disease, physiological disabilities (hearing, vision, etc.), physical fatigue, etc.;

– physical or mental limitations, which render an individual unable to successfully complete a task: insufficient reaction time, lack of training, lack of physical ability and lack of intellectual ability;

– aspects linked to fitness to work: respect of rest periods, limitations concerning the consumption of alcohol, etc.

The state of the working environment constitutes another category of preconditions. This environment includes:

– the physical environment, which may generate problems linked to noise, heat, etc.;

– the technical environment with which the operator interacts to acquire information and execute actions: causes linked to this aspect include, for example, problems with user–machine interfaces or design faults in machinery;

– the human environment, which is linked to causes such as insufficient communication, inefficient sharing of knowledge or abilities, poor coordination and/or synchronization of tasks or interference from external parties.

3.6.1.3. *Failures in hierarchical supervision*

The next level concerns failures in hierarchical supervision, which may be grouped into several categories:

– Inadequate supervision: the supervisor fails to provide direction, training, leadership, supervision or adequate incentives to carry out tasks in an efficient and safe manner.

– An inappropriate plan of operations: this corresponds to operations that would be unacceptable in normal operating conditions, even if such actions may be acceptable in exceptional circumstances, such as in cases of emergency.

– Inability to correct a known problem: this includes cases where problems are known to supervisors, but where no measures are taken.

– Violation of supervision: cases where existing rules and regulations are deliberately ignored by a supervisor.

3.6.1.4. *Organizational failures*

The organizational failure level is divided into three categories:

– Resource management failure: this concerns the organizational level of decision making concerning the assignment and maintenance of organizational assets, such as human resources, financial resources and equipment.

– Failure in the organizational environment, which generates a working atmosphere within the organization, which can make faults more likely to occur (e.g. structure, policies and culture).

– Failure in the operational process: this concerns organizational decisions and the rules that govern daily activities within an organization (e.g. operations and monitoring procedures).

This taxonomy of errors and causes of errors guides the analyst in the identification of risks. It remains, however, a simplified description of the link between human, technical and organizational aspects.

3.6.2. *Tripod classification of organizational failures*

The Tripod method [SCH 09] offers a classification of organizational failures, constructed through the *a posteriori* analysis of a certain number of accidents and incidents and the study of different audit reports. The different latent failures have been grouped into a limited number of basic risk factors, or BRFs. The 11 Basic Risk Factor (BRF) categories are shown in Figure 3.5.

BRF	Description
DE Design	Quality of the design of the workplace, equipment and materials in ergonomic and operational terms
HW Hardware	Quality, state, availability of tools and equipment
MN Maintenance	Planning, quality and implementation of maintenance and repair activities
EC Error enforcing conditions	Quality of the physical work environment: luminosity, noise levels, temperature, air. Quality of the psychological work environment: motivation, stress, work satisfaction, attitude, working ambiance with colleagues
PR Procedures	Availability, quality, up-to-dateness, relevance of operating modes, usefulness of procedures, working instructions
TR Training	Planning, coordination and effectiveness of training Professional experience of personnel
CO Communication	Quality of communication between employees, services and sites, in terms of the availability of resources and the effectiveness of communication channels
IG Incompatible goals	Way in which safety is managed in relation to other objectives, such as production or financial goals
OR Organization	Quality of workplace organization, task distribution, definition of responsibilities
DF Defenses	Availability and effectiveness of protection equipment and protective measures
OP Order and cleanliness	Orderliness of different workspaces

Figure 3.5. *Organizational factors used in Tripod*

Chapter 4

The Risk Management Process (ISO31000)

4.1. Presentation

Risk management may be defined as a set of coordinated activities with the aim of directing and guiding a organization in relation to risk. The risk management process used to implement these activities consists of the systematic application of principles, policies, procedures and practices for risk identification, analysis, evaluation and treatment tasks. It also includes aspects of communication, establishment, consulting, monitoring and risk review.

During the 1990s, the need for a global and improved risk management approach became apparent, and a number of different standards have since been suggested. The first version of the AS/NZS 4630 (*Australian/New Zealand Standard*) risk management standard was published in 1995, and is considered to be the first risk management standard. This document provides a full description of the risk management process. It was followed by a number of national and domain-specific standards, such as the IEC 61508 standard.

The AS/NZS 4630 standard was extended in 1999, with the notable addition of a section on communication. A new edition, published in 2004, took account of feedback from the past 10 years of experience. This version was proposed to the International Organization for Standardization (ISO) in the same year,

54   Risk Analysis

and led to the creation of the ISO31000 standard in 2009. This new standard replaced the AS/NZS 4630 standard.

The general idea contained in these approaches consists of defining:

– risk management principles, which define the reasons for the implementation of risk management;

– an operational framework, which defines how risk management should be integrated into the strategy of an organization;

– a management process, which defines how risk management should be carried out in a concrete manner.

Figure 4.1. *Construction of the ISO31000 standard*

The ISO31000 standard was established as a result of an ISO standardization procedure that ran from 2005 to 2009. During this period, the standard was established as a consensus based on a variety of experiences. The ISO31000 is a standard, and not a set of regulations; therefore, it needs to be adopted on a voluntary basis. The creation period also allowed for the integration of different points of view from different domains. The ISO31000 standard can therefore be seen as a "blanket" or "hat" norm (Figure 4.2), used to establish coordination between different sectors of activity by providing a framework and a common vocabulary. The standard is highly abstract, meaning that it can be implemented by organizations of any size, operating in any sector and with risks of any type (individuals, companies, collectivities, governments, non-governmental organizations (NGOs), etc.). Its aim is not to establish uniformity in practice, but to harmonize approaches in terms of principles and processes. This harmonization promotes the development of training and renders the implementation of risk management more effective.

To adapt to the various issues involved in risk management, the ISO31000 standard uses the very general definition of risk presented in Chapter 2, that is the effect of uncertainty on objectives.

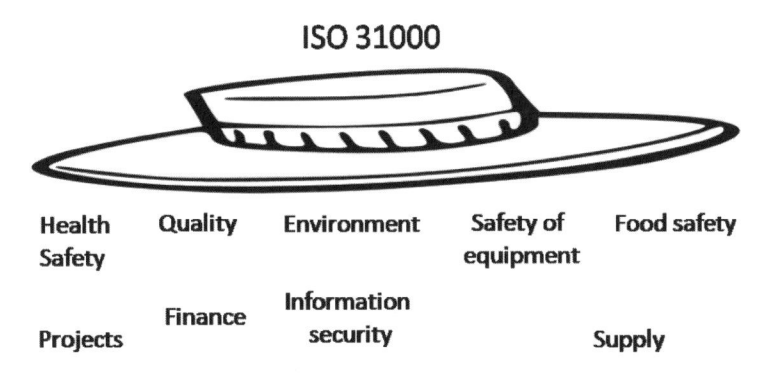

ISO 31000

| Health Safety | Quality | Environment | Safety of equipment | Food safety |
| Projects | Finance | Information security | | Supply |

Figure 4.2. *"Umbrella standard" ISO31000*

4.2. ISO31000 standard

4.2.1. *Basic principles*

The ISO31000 standard defines 11 principles that a organization should follow in a risk management framework for this management approach to be effective:

1) Risk management creates and preserves value. This statement highlights the fact that the set of risk management activities should provide a tangible contribution to the attainment of objectives and to improve the performance of the organization.

2) Risk management is an integral part of organizational processes. It must be integrated into the existing management system, including operational level, and must not be developed in parallel.

3) Risk management is a part of the decision-making process. The risk level constitutes an important parameter in decision making.

4) Risk management explicitly deals with uncertainty. This uncertainty may concern sources of risk, in the case of emerging risks. The approach

should take this lack of knowledge into account in an explicit manner. Uncertainty may also concern analytical methods and the way in which risks are treated (safety barriers). The impact of uncertainty should be taken into account.

5) Risk management is systematic, structured and timely in order to ensure that it produces effective, relevant, coherent and reliable results.

6) Risk management is based on the best available information, both for the system in question and the data and models being used.

7) Risk management is tailored in terms of hazard levels to the capacities of the company and to the available human and financial resources.

8) Risk management takes human and cultural factors into account. In Chapter 3, we saw that a significant proportion of the causes of accidents depends on these factors.

9) Risk management is transparent and inclusive. It is important to involve all interested parties, both internal and external, in the risk management process, notably to define what is acceptable and what is not acceptable.

10) Risk management should be dynamic, iterative and responsive to change. It must adapt to changes in the business context and the evolution of risks.

11) Risk management facilitates continuous improvement and enhancement of the organization.

4.2.2. *The organizational framework*

The purpose of the organizational framework is to integrate risk management activities into the organizational management structure. This framework is defined by a process involving the implementation of risk management processes and their continual improvement (Figure 4.4). It involves a *Plan, Do, Check, Act* (PDCA) type cycle, following a clear definition of a mandate and commitment based on risk management principles. Task 4.4 of the implementation process is based on the risk management process.

The purpose of this framework is not to give a prescribed management system, but to assist organizations in integrating risk management into their overall management system. The PDCA structure is similar to that used in other company management systems, such as:

- ISO9000: quality;
- ISO14000: environment;
- OHSAS18001: health and safety;
- ISO27000: information security;
- ISO26000: social responsibility;
- ISO28000: supply chain security.

Principles (Clause 3)
a) Creates value
b) Integral part of organizational processes
c) Part of decision making
d) Explicitly addresses uncertainty
e) Systematic, structured and timely
f) Based on the best available information
g) Tailored
h) Takes human and cultural factors into account
i) Dynamic, iterative and responsive to change
k) Facilitates continual improvement and enhancement of the organization

Organizational framework (Clause 4)
- 4.2 Mandate & commitment
- 4.3 Design of organizational framework
- 4.6 Continual improvement of the framework
- 4.4 Implementation of risk management
- 4.5 Monitoring and review of the framework

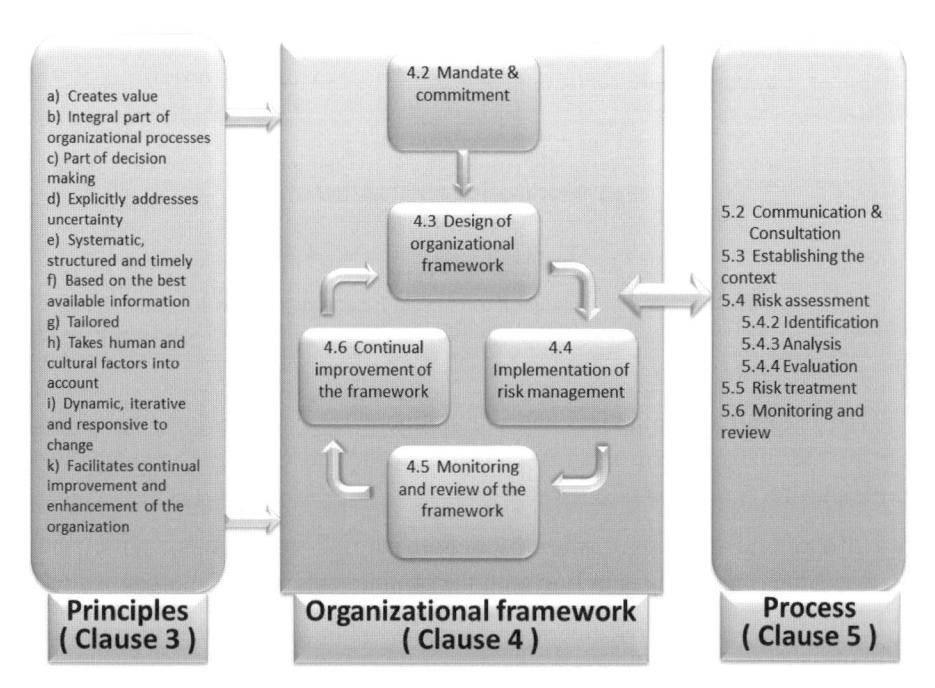

Process (Clause 5)
- 5.2 Communication & Consultation
- 5.3 Establishing the context
- 5.4 Risk assessment
  - 5.4.2 Identification
  - 5.4.3 Analysis
  - 5.4.4 Evaluation
- 5.5 Risk treatment
- 5.6 Monitoring and review

Figure 4.3. *Organizational framework for risk management (ISO31000 standard)*

However, the ISO31000 standard defines an organizational framework for the design of a management system, rather than a framework for the implementation of risk management. This aspect is not defined in detail, allowing greater flexibility at company level. Moreover, the ISO31000 standard is not intended to serve as a basis for certification.

58   Risk Analysis

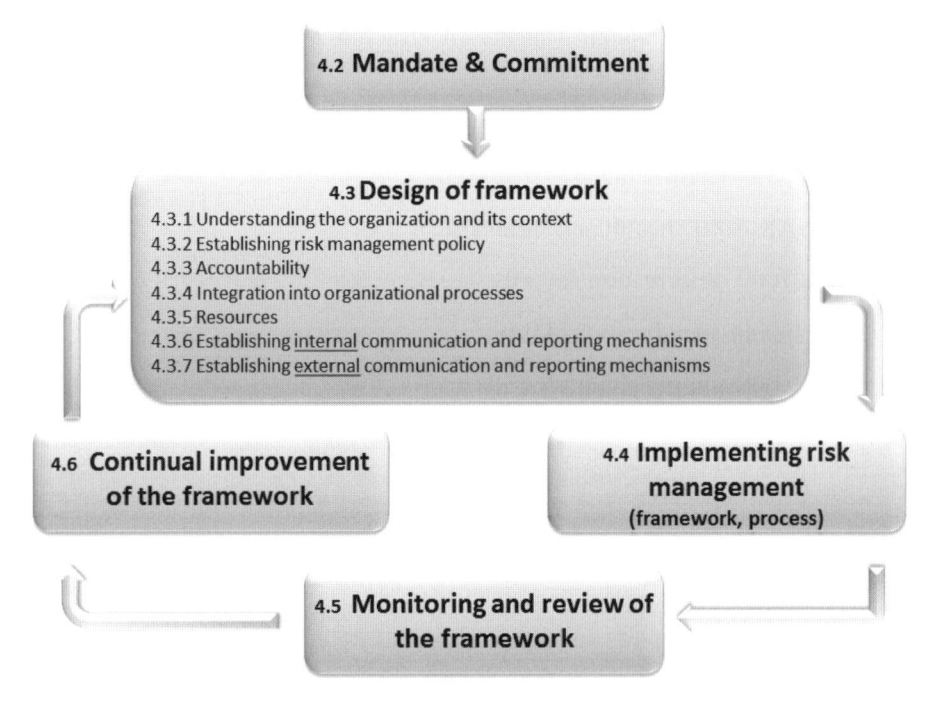

```
                    ┌─────────────────────────────┐
                    │  4.2 Mandate & Commitment   │
                    └──────────────┬──────────────┘
                                   ▼
        ┌──────────────────────────────────────────────────────┐
        │              4.3 Design of framework                 │
        │  4.3.1 Understanding the organization and its context│
        │  4.3.2 Establishing risk management policy           │
        │  4.3.3 Accountability                                │
        │  4.3.4 Integration into organizational processes     │
        │  4.3.5 Resources                                     │
        │  4.3.6 Establishing internal communication and reporting mechanisms │
        │  4.3.7 Establishing external communication and reporting mechanisms │
        └──────────────────────────────────────────────────────┘

  ┌─────────────────────────┐              ┌─────────────────────────┐
  │ 4.6 Continual improvement│              │ 4.4 Implementing risk   │
  │    of the framework     │              │      management         │
  │                         │              │  (framework, process)   │
  └─────────────────────────┘              └─────────────────────────┘

                    ┌─────────────────────────────┐
                    │ 4.5 Monitoring and review of│
                    │       the framework         │
                    └─────────────────────────────┘
```

Figure 4.4. *Organizational framework for risk management (ISO31000 standard)*

The preliminary "mandate and commitment" phase is essentially carried out by the management of an organization, who must use risk management principles for a number of purposes:

– To demonstrate implication by defining a risk management policy and ensuring that the policy is in line with the broader culture of the organization.

– To define risk management performance indicators in accordance with organizational performance indicators.

– To assign responsibilities to appropriate levels so that risk management is carried out at all levels of management.

– To ensure that the necessary resources are allocated.

– To ensure legal and regulatory compliance.

– To communicate with all interested parties concerning risk management.

– To ensure that the organizational framework remains suitable.

The task of designing the organizational framework itself (*plan*) includes the following aspects:

– Understanding the organization and its context: this includes an evaluation of the external environment (social, cultural, legal, political, regulatory, financial, etc.), factors with impact on the objectives of the organization and relationships between the organization and external interested parties. Similarly, the internal environment must be evaluated, considering aspects such as direction, organization, objectives, culture, etc.;

– Establishing a risk management policy: this involves specifying objectives for the organization in this domain, explaining motivations, defining links between the risk management policy and other objectives, defining how conflicts of interest should be treated (e.g. how to reconcile the need for consensus concerning industrial risks with the need for discretion to avoid the risk of malicious activity), establishing ways to measure performance in risk management, presenting the commitments of the organization in terms of provision of resources and establishing accountability in terms of risk management, then communicating in relation to this policy.

– Defining accountability (in the sense of active responsibility and not in a legal context): this aspect ensures that responsibility, authority and abilities are established in terms of risk management. The standard introduces the notion of risk ownership, where a nominated individual is in charge of a risk and has the authority to manage it. This notion is controversial as it does not allow for collective responsibility.

– Providing adequate resources: different resources must be allocated, including human resources, methods, tools, procedures, information management systems and training programs.

– Integrating the framework into organizational processes: the aim of this aspect is to integrate risk management into all processes within the organization. It is useful to establish a risk management plan on an organizational scale in order to ensure that the risk management policy is implemented and that these practices are integrated into all processes within the organization. A plan should specify an approach, management components (procedures, practices, assignment of responsibility, planning) and associated resources.

– Establishing internal communication and reporting mechanisms: this aspect includes the establishment of communications concerning components of, and modifications to, the organizational framework, the establishment of a consultation process with interested parties and the creation of reports on the effectiveness of the risk management process.

– Establishing external communication and reporting mechanisms: this aspect concerns the creation of a communications plan with and for external stakeholders, including the way in which they participate in the risk management process, the creation of external reports in accordance with legal and regulatory requirements, the establishment of feedback mechanisms for communications and consultation and communication methods in crisis management situations.

The implementation (*Do*) of risk management involves two aspects:

– The implementation of the organizational framework. Once the framework has been defined, the implementation consists of transforming theory into practice and making the risk management system efficient enough. More precisely, this involves ensuring the risk management process is understood by risk owners (through high-quality communications and training) and that the risk is effectively managed. An implementation schedule and strategy should be defined. The management process, described below, must be applied and legal obligations must be respected. Moreover, the decisions taken by the organization must be coherent with respect to the findings of the risk management process. Finally, the organization must ensure that the organizational framework remains appropriate through consultation with stakeholders.

– The implementation of the risk management process, discussed in the following sections, which is applied to all functions at all relevant levels of the organization and which corresponds to the classic activities involved in what is commonly known as risk mastery.

The monitoring and review phase (*Check*) of the organizational framework consists of:

– establishing key performance indicators (KPIs) to measure the performance of risk management and evaluate progress and deviations in relation to the risk management plan;

– verifying the suitability of the internal and external contexts established in phase 3.2;

– producing reports on risks, the advancement of the plan and the way in which the management policy is followed;

– verifying the effectiveness of the organizational risk management framework.

The final stage (*Act*) is the continual improvement of the operational framework, based on the results of the previous stage. These actions may concern improvements to the framework itself, to the policy or to the risk management plan.

4.3. Implementation: the risk management process

The concrete implementation of risk management is carried out following the generic risk management process shown in Figure 4.6. It involves classic risk assessment activities (identification, analysis and evaluation) and a risk treatment phase. This approach was formalized in the ISO/CEI 51:1998 guide (Figure 4.5), with the addition of three other activities:

– establishment of context;

– communications and consultation;

– monitoring and review;

4.3.1. *Establishing the context*

The context establishment process precedes risk assessment activities, and aims to define:

– characteristics of the external environment in which risk management takes place;

– characteristics of the internal environment;

– characteristics of the risk management process;

– criteria used to characterize the importance of a risk.

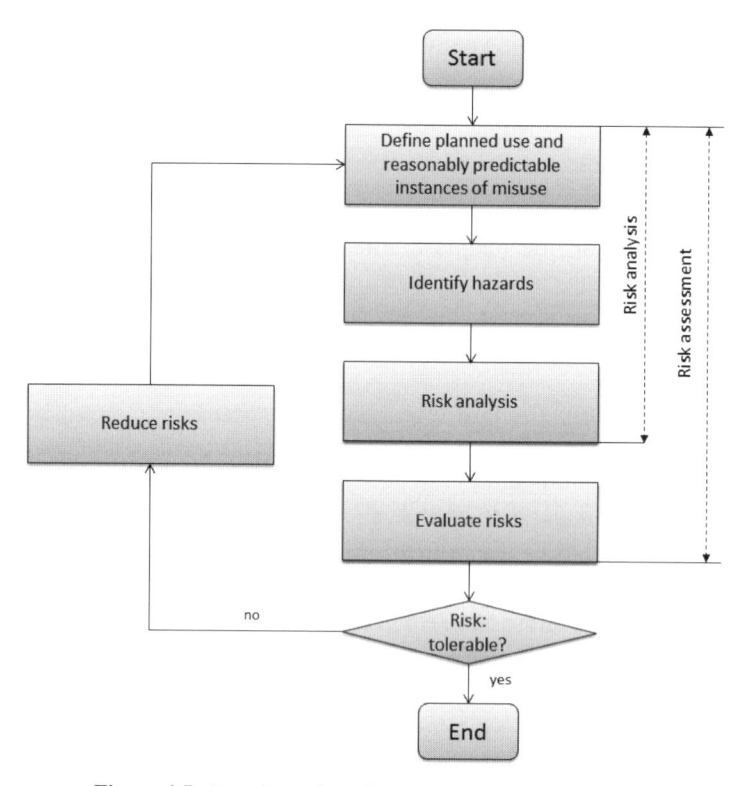

Figure 4.5. *Iterative risk evaluation and reduction process*

The first two points should already have been defined in the organizational risk management framework. However, it is useful to examine them in detail to evaluate their relevance in the context of the process (domain of activity, type of risk, subgroup within the organization, etc.).

From a practical standpoint, the external context allows us to establish aspects including:

– the regulatory and legal context;

– the social context;

– a list of stakeholders and their points of view.

The Risk Management Process (ISO31000)

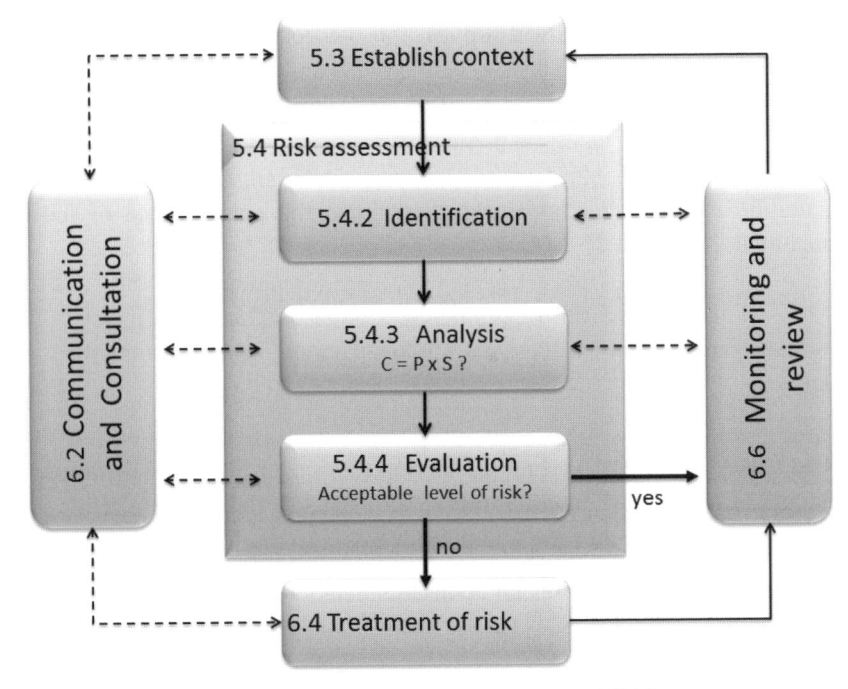

Figure 4.6. *Risk management process (ISO31000)*

The characteristics of the risk management process are used to define the framework for operations, and include:

– goals and objectives: in cases involving health or safety, the goal is always to reduce the risk to an acceptable level. For other types of risk, objectives may be expressed in a different manner;

– definition of responsibilities in relation to the risk management process: who is responsible for what?

– definition of the domain of application, the degree and the extent of risk management activities, what is included or excluded – for instance which risks are taken into account?

– definition of the object of the risk management process (activities, processes, functions, projects, etc.) in terms of time and place;

– definition of the relationship of this object to other entities in the organization;

– definition of risk assessment methods: identification, analysis and evaluation methods;

– definition of the method used to evaluate the performance of the risk management process;

– identification of decisions that need to be taken following risk assessment;

– characterization of necessary studies (nature, extent, objectives and resources).

The final aspect to define concerns the criteria that will allow an organization to measure the importance of a risk. To define these criteria, the standard recommends that consideration is given to the following factors:

– nature and type of causes and consequences;

– method used to define likelihood;

– scale of likelihood and/or importance of consequences;

– method used to determine the level of risk;

– method used to take account of the opinions of stakeholders;

– level at which a risk may be considered to be acceptable or tolerable;

– possible consideration of combinations of several risks and, where this is done, the method used.

Establishing the context is a stage in formalizing the definition of the framework of the study: it allows us to define the object of study, its interactions with its environment, the nature of the risks being studied and the types of consequences to take into account. We also determine the methods chosen for the identification and analysis of risks, scales of probability and severity, the risk matrix and thresholds of acceptability. This stage is carried out prior to using these parameters in other phases of study, clarifying the precise responsibilities of the different entities involved.

4.3.2. Risk assessment

4.3.2.1. Identification

This step consists of identifying risks and describing them in detail. We identify sources of risks, targets in the different domains concerned by the studies, events leading to the occurrence of risk and potential causes and consequences. Risks are described in full in text and/or scenario form. An exhaustive identification process is important, as any risk not identified at this stage will not be treated later in the process.

4.3.2.2. Analysis

The aim of the analysis is to estimate the importance of a risk, that is its level, by estimating:

– the likelihood of a risk in consideration of credible causes;

– the severity of direct or indirect consequences of the event that characterizes the risk.

Before this estimation is carried out, we must analyze the risk to obtain the data needed to determine which causes to take into account, to evaluate the importance of consequences and to evaluate their likelihood. A risk level is then determined by combining the likelihood and the severity of consequences. This combination is generally carried out using a risk matrix (see Chapter 6). Qualitative, quantitative or semi-quantitative approaches may be used.

It is sometimes difficult to determine this risk level. The ISO31000 standard specifies that the level of trust placed in the determination of a risk level, sensitivity to hypotheses and preconditions should be taken into account in the analytical process and effectively be communicated to the decision makers and, where necessary, to the stakeholders. Uncertainties, limitations of modelings or differences in expert opinions should be mentioned. This aspect has not been widely developed in practice, although it is beginning to be treated in a systematic manner, notably in the case of emerging risks.

4.3.2.3. Evaluation

The analytical process produces a result characterizing the level of risk and the degree of trust placed in this evaluation. The aim of the evaluation process is to assist decision makers to identify risks requiring treatment, and to establish priorities for the implementation of these treatments.

This evaluation consists of comparing the risk level to decision criteria defined during the context establishment phase in order to study the need for treatment. The notion of tolerable risk, as defined in [ISO 99], is not cited explicitly; however, the need for treatment is defined based on:

– the results of comparison, used to determine whether a risk is tolerable, intolerable, as low as reasonably practicable (ALARP), etc.;

– the risk tolerance of various stakeholders and legal obligations;

– existing action plans.

4.3.3. *Treatment of risk*

Once the risk has been evaluated, we have four possible options (the 4 Ts):

– Treat the risk, with the aim of reducing it:

    - by eliminating the source, or removing the target from its field of action;

    - by modifying the importance of consequences;

    - by modifying the likelihood of consequences.

– Terminate the risk-generating activity: in this case, the organization refuses to accept the risk, and the decision is made not to continue (or not to begin) the activity that generates the risk.

– Transfer the risk, or share the risk with (an)other party (parties), for example using insurance.

– Tolerate the risk: in this case, the organization decides to take a risk in order to pursue an opportunity (case of risks with positive consequences) or to accept a risk based on detailed argumentation (case of risks with negative consequences, such as health and safety risks).

This is an iterative process: once the risk reduction methods have been established, we must verify whether or not the residual risks are tolerable, and, if this is not the case, we must implement additional measures. It is also important to check that these efficient techniques do not themselves create new risks, and that they are effective, that is the possibility of failure does not generate an excessively high level of risk.

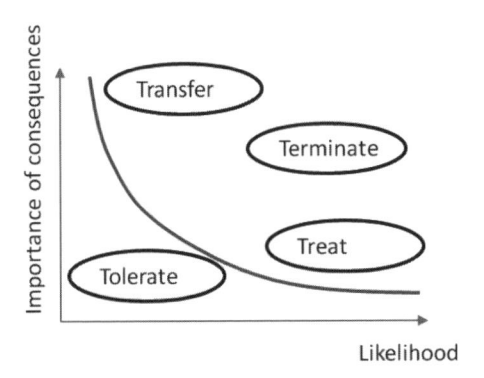

Figure 4.7. *Treatment of risk (four Ts)*

The choice of a risk treatment action involves comparing the costs and difficulties of implementing different possibilities with regard to the obtained advantages, taking account of legal and regulatory obligations, and other requirements, such as social responsibility. The ISO31000 standard indicates that decisions must also take account of risks for which treatment cannot be justified from an economic standpoint, as in the case of major risks that are very unlikely to occur.

Treatment actions are organized into a treatment plan, giving a list of actions to implement, the motivations for these actions, the person or persons responsible, the resources needed for the actions and an implementation schedule.

As part of a continual improvement approach, this plan must also specify the way in which the performance of treatment actions should be measured, alongside monitoring and reporting requirements.

4.3.4. *Communication and consultation*

The ISO31000 standard formalizes the communication and consultation task and explicitly sets out links with different elements of the process. Consultation and communication plans must be established from the very beginning of the process and must cover the risk itself, its causes, its consequences and the measures taken. A team-based consultative approach is used to:

– help to correctly define the context and take account of the interests of all parties involved;

– ensure that risks have been identified correctly;

– carry out analysis using different domains of expertise;

– carry out evaluation including different viewpoints;

– promote adhesion and support for the treatment plan;

– produce a relevant communication and consultation plan.

This stakeholder consultation phase is important as it allows all of the entities involved to express their perception of a risk.

4.3.5. *Monitoring and review*

The aim of this stage is to monitor the performance of the risk treatment process. It should allow us to:

– ensure that efficient techniques are in place, are effective and well performing;

– analyze and learn from different past events (accidents, near-accidents and incidents);

– detect changes in the context;

– identify emergent risks.

This step may be carried out through audits and reviews on a periodic or occasional basis.

4.3.6. *Risk evaluation methods*

The ISO31010 standard is closely linked to the ISO31000 standard. It provides a guide to the choice and application of risk assessment methods. Table 4.1 presents a list of the methods described in this standard, most of which are presented in this book. The implementation of these methods within the framework of the ISO31000 standard allows a more systematic approach with an explicit definition of working parameters, integrating risk management into the management activities of the organization.

1	Brainstorming
2	Structured or semi-structured interviews
3	Delphi
4	Checklists
5	Preliminary Hazard Analysis
6	HAZOP
7	HACCP
8	Environmental risk assessment
9	Structured What-if ?
10	Scenarios analysis
11	Business impact analysis
12	Root cause analysis
13	FMEA (AMDEC)
14	Fault tree
15	Event tree
16	Cause-consequence analysis
17	Cause-and-effect analysis
18	LOPA
19	Decision tree
20	HRA
21	Bowtie Analysis
22	RCM
23	Sneak circuit analysis
24	Markov analysis
25	Monte Carlo simulation
26	Bayesian statistics and Bayes nets
26	FN curves
28	Risk indices
29	Consequence/ Probability matrix
30	Cost benefit analysis
31	Multi-criteria decision analysis (MCDA)

Table 4.1. *Methods presented in the ISO31010 standard*

Part 2
Knowledge Representation

Chapter 5

Modeling Risk

5.1. Introduction

During the risk assessment phase, starting with the identification process, we need to describe identified scenarios in detail. More specifically, we need to describe the source of a risk, the event characterizing the occurrence of the risk, its causes and its consequences. This allows us to better understand the conditions in which the risk occurs and facilitates analysis and evaluation of the importance of the risk. A number of different approaches to this description or modeling of risk have been put forward. In this chapter, we will present the most widely used models. The aims of these modelings can include:

– investigation after an accident;

– analysis and comprehension of the conditions in which a risk occurs;

– evaluation of importance or quantification of a risk level.

In the following chapters, we will see how these models may be constructed during the risk analysis process and how they may be used.

These models may be grouped into three categories:

– The first category of model represents degradation flows, that is the undesirable action of a source generating a hazardous phenomenon for a target.

– The second category is made up of models using event graphs, which describe the occurrence of a risk as a succession of events, using a causality-based representation leading from the initiating event to final damage.

– The third category, dynamic models, is used to describe the evolution of a system over time, from a normal to an abnormal state.

In this chapter, we will provide a relatively informal presentation of these different types of model. Our aim is to give an overview of different approaches and to show how they may complement each other. A more detailed discussion of these modelings will be given in Part 3 of this book.

5.2. Degradation flow models

5.2.1. *Source–target model*

One widely used approach to modeling risk consists of representing a degradation flow moving from a hazard source, which generates the hazardous phenomenon, toward a target that suffers the effects of this hazardous phenomenon (Figure 5.1). This representation was developed based on the concepts put forward by Haddon [HAD 73]. It mainly concerns risks generated by energy sources, defined as "any machine, mechanism or process which contains energy which may potentially be released and cause damage to a potential target". The hazard source may also be a source of toxicity, which propagates toward a target. This model has the same basic structure as the system dysfunctions analysis methodology (SDAM) model used in the SORAM method (Chapter 11).

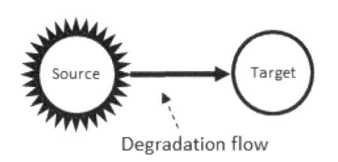

Figure 5.1. *Source–target modeling*

The first barrier-based prevention approaches were developed using this representation. In this approach, a barrier is defined as any means of preventing a source from creating damage to the target. It may act by protecting the target or prevent the target and the source from meeting. Barriers can include physical barriers, imposed distances or procedures. They may be represented at source or target level.

This approach is well suited to modeling risks caused by an energy source that becomes uncontrollable, or those generated by a source of hazardous products that may be released from their confined container. It is useful in describing the propagation of degradation flows and analyzing barriers, but it does have certain limitations. The root causes of the degradation of the target by the source, for instance, are not shown and the source is only represented in its abnormal state. As a result, certain preventive barriers, which act by preventing the source from attaining an abnormal state, cannot be represented.

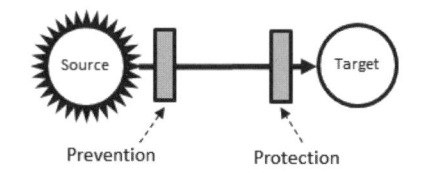

Figure 5.2. *Source–target modeling with barriers*

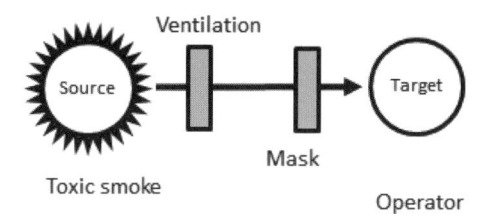

Figure 5.3. *Example of a source–target modeling with barriers*

For this reason, the model described above is often supplemented with a hazard triangle model (Figure 5.4), where the initiating event at the root of the degrading action is also represented.

5.2.2. *Reason's model*

Reason's model [REA 97] also represents the propagation of a degradation flow (Figure 5.5), and has already been discussed in section 3.6.1. Defenses, barriers and safety mechanisms may be represented as slices of Swiss cheese,

76   Risk Analysis

with holes representing failures in each level of defense, and thus the causes of damage. The model shows different types of barrier (Figure 5.5).

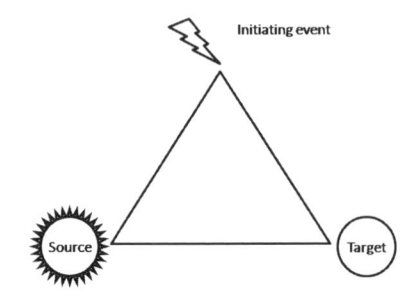

Figure 5.4. *Hazard triangle*

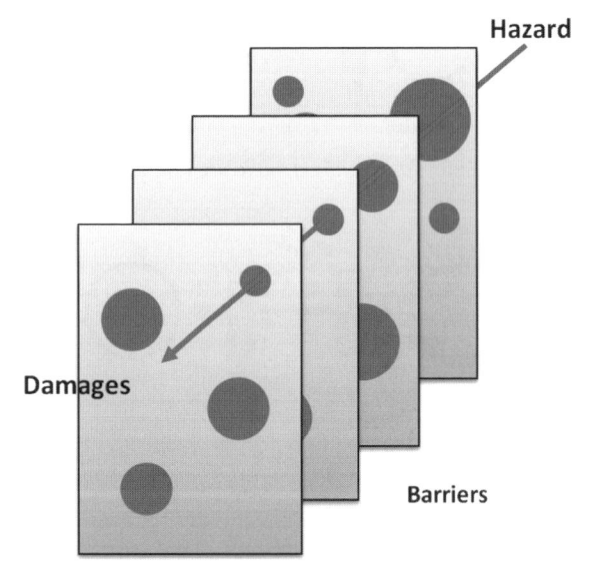

Figure 5.5. *Reason's model*

Reason's model is important from a conceptual perspective, as it highlights the complexity of cause-effect relationships, showing that a single cause is not sufficient; a succession of failures in barriers is necessary for a latent cause to produce damage.

NOTE 5.1.– This model is sometimes considered to describe the occurrence of an accident in a similar way to the development of a disease, and is thus said to be an epidemiological model. It effectively considers that several pathogenic factors are present in the system, which combine to create a possible pathway toward an accident through the barriers present in the system (Figure 5.5). We therefore need to seek out and destroy these pathogenic agents, which reduce the effectiveness of barriers.

5.2.3. *From source–target to causal modeling*

Source–target models only provide a static view of a situation. This method does not allow us to represent the switch from a normal state to a degraded state, in which risk may occur, to a state where the hazardous phenomenon effectively damages the target. To represent these evolutions, we need to use a representation that shows the connections between causes and consequences.

Figure 5.1 shows this sequence in connection with the degradation flow model: initially, the source is in a normal state, in which damage cannot occur. Following a parameter deviation or a fault, the source moves into a hazardous state. In this state, the hazardous phenomenon can take place if a initiating factor appears, and, if the target is present, can generate damage.

EXAMPLE 5.1.– Let us consider a piece of electronic equipment as a hazard source. Isolating material is degraded as the result of wear and tear, and the equipment ceases to be isolated. Thus, it becomes a hazard source. During contact with the equipment – the initiating event – a user receives an electric shock, which results in the accident. It is important to clearly distinguish between arrows representing causality relationships and those describing degradation flows, shown inside boxes in Figure 5.6. In an event-based causal modeling, each place in the graph is known as an "event", and we obtain an event graph, retaining causality connections. The degradation flows are no longer shown.

5.3. Causal modeling

The representation of the chain of events leading to an event characterizing a risk, followed by the description of the related consequences using a graph, constitutes an interesting modeling approach. The formalisms presented in this section use this idea; some of them also allow the representation of barriers.

78   Risk Analysis

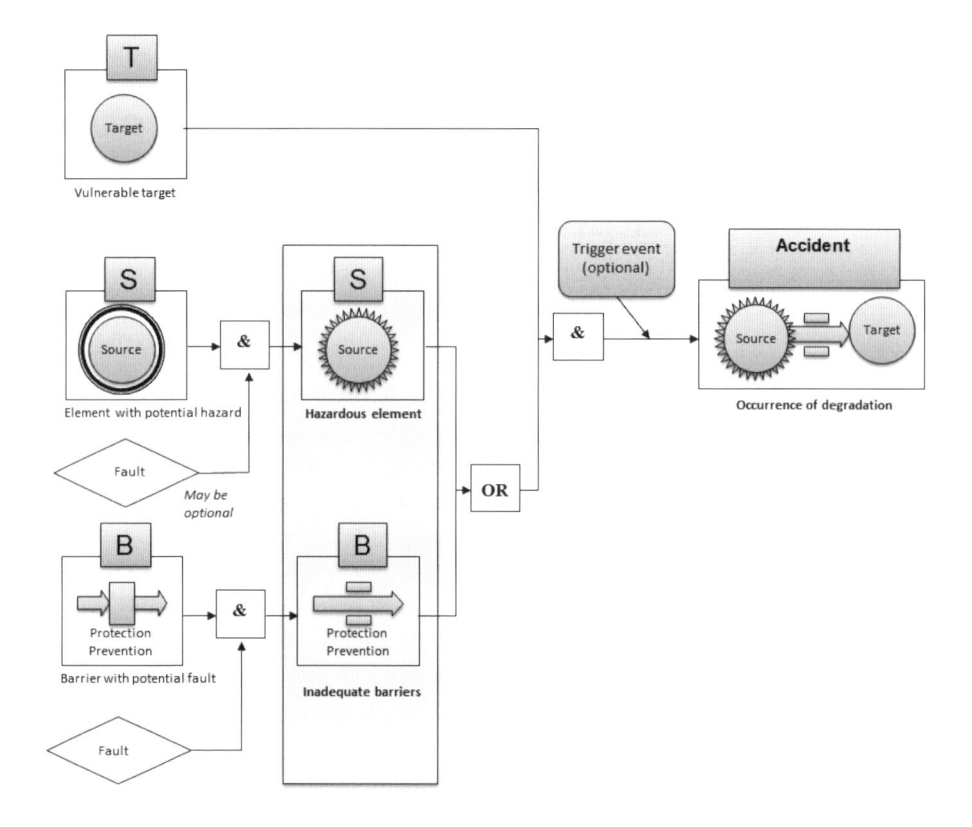

Figure 5.6. *From the source–target model to a causal model*

5.3.1. *Fishbone cause–effect diagram*

The first type of representation is based on an Ishikawa diagram, also known as a fishbone diagram. This diagram (Figure 5.7) represents an effect, in this case the occurrence of a hazardous phenomenon or of damage, as the consequence of a certain number of causes falling into five or six categories: causes linked to materials, machines, methods, man power, the milieu (also called mother nature) which is the context, and, in certain cases, measures.

This type of approach is well suited to risks concerning systems with components of varying natures. However, it is not particularly formalized and cannot be applied to highly complex systems, as it cannot easily combine sequences of multiple diagrams. It is better to limit the use of this method to the creation of exhaustive inventories of the causes of a given effect, and to use another approach to carry out more detailed analysis or to quantify the likelihood of the risk.

EXAMPLE 5.2.– This method can be applied in analyzing the causes of a risk of a manufacturing fault by a station composed of a machine and an operator, receiving an input of raw materials and producing a manufactured piece. These causes may be linked to:

– materials: faults in the incoming element, poor quality, wrong lubricant used in the process, etc.;

– the machine: use of the wrong settings, poor maintenance, etc.;

– methods: unsuitable methods, poorly described or incomplete procedures, etc.;

– the workforce (operator and management): badly trained operator, non-respect of procedures, imprecise instructions, etc.;

– the context: noisy environment, excessive pace of work, etc.

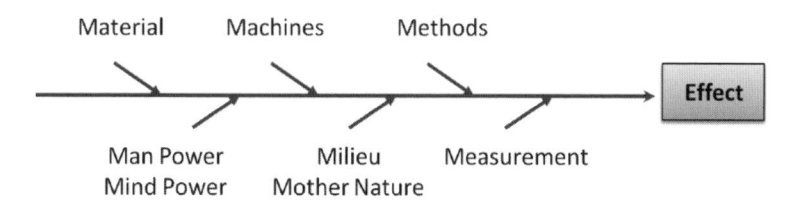

Figure 5.7. *Fishbone model*

5.3.2. *Causal trees*

The causal tree model may be seen as an evolution of the Ishikawa diagram, in which causes are not grouped into categories, allowing a more flexible representation. This model is often used as part of an analytical procedure to identify root causes after an accident.

This method was designed by the *Institut National de Recherche et de Securité* (INRS) in the 1970s. The "tree" is oriented from left to right (Figure 5.8). Each leaf corresponds to a fact. The element which is furthest to the right is the final fact, generally the effect that we wish to explain. The causes of each fact are linked to the leaf as input. The connectors are implicitly considered to be AND connectors, meaning that a fact occurs if all of its causes have occurred. Other connector types, such as OR, do not exist in this modeling as the causality tree aims to explain a fact that has already taken place. The method used to construct the tree is given in Appendix 2.

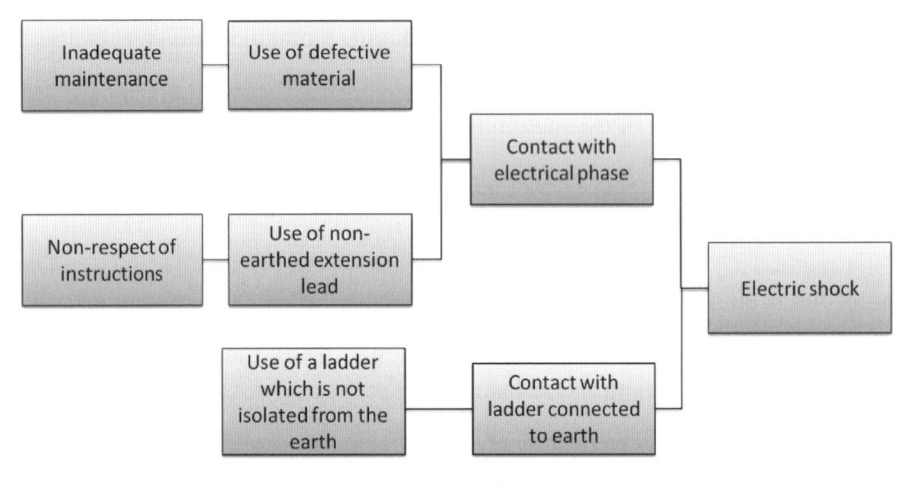

Figure 5.8. *Example of a causal tree*

5.3.3. *Fault tree*

A fault tree (Figure 5.9) is a representation of the causes of an undesirable event, generally that represents the occurrence of a risk, in tree form. Events are linked by OR and AND connectors, signifying, respectively, that the output event occurs if one of the input events occurs, or if all of the input events occur. The difference between fault tree and causal tree models is that the fault tree can represent all of the possible combinations leading to the undesirable event, and, for this reason, different types of logical connector may be used. Notably, fault tree models make use of the OR connector that is not present in a causal tree model. This approach is more highly formalized, allowing for algorithmic treatment to identify combinations of events leading to the occurrence of the undesirable event.

Moreover, the fault tree model is often represented vertically, from top to bottom, rather than from left to right (Figure 5.9). This approach is discussed in detail in Chapter 12.

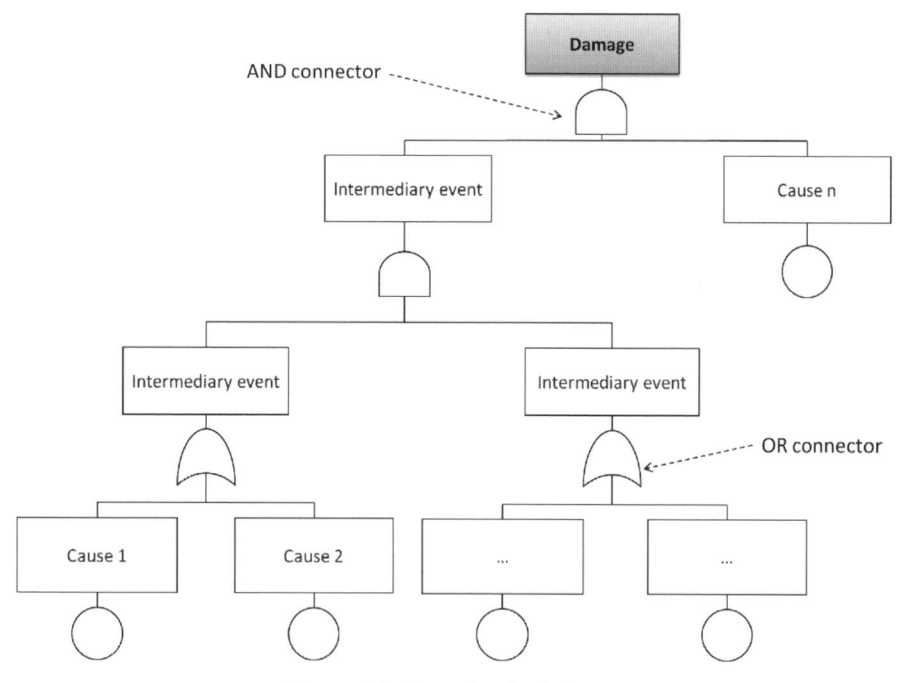

Figure 5.9. *Example of a fault tree*

The fault tree is a tool that may be used for quantification. Using the laws of probability, it is possible to evaluate the probability of an event as output from a gate based on the probability of the input events, supposing that these are independent.

EXAMPLE 5.3.– Let us consider an AND gate with two independent input events, A and B, and an output event, S. If $P(A) = 0.1$ and $P(B) = 0.02$, we can deduce that $P(S) = P(A)P(B) = 0.002$.

A fault tree may be constructed directly by considering the causes of an undesirable event, then identifying the causes of these causes, and so on until the "root" causes have been identified. However, in the context of risk management processes, fault trees are constructed in a more systematic fashion, following the risk identification phase. This mode of operation is discussed in Chapter 12.

5.3.4. *Consequence or event trees*

Fault trees are used to represent the causes of an undesirable event. Consequence trees, as their name suggests, are used to represent the consequences of an event. They are constructed from left to right, starting from the undesirable event associated with, or initiating, the risk. For each envisaged consequence, a node is added on the right, and the arc is labeled with the associated probability of occurrence. When consequences are mutually exclusive and their number is limited to two, we obtain a binary tree (Figure 5.10) of the success–failure type, which is, in particular, suitable for analyzing safety measures: one branch corresponds to consequences when this measure operates successfully, the other branch corresponds to cases where the measure fails. As the two alternatives are mutually exclusive, if the first option has a probability p, the second option will have a probability of $1 - p$.

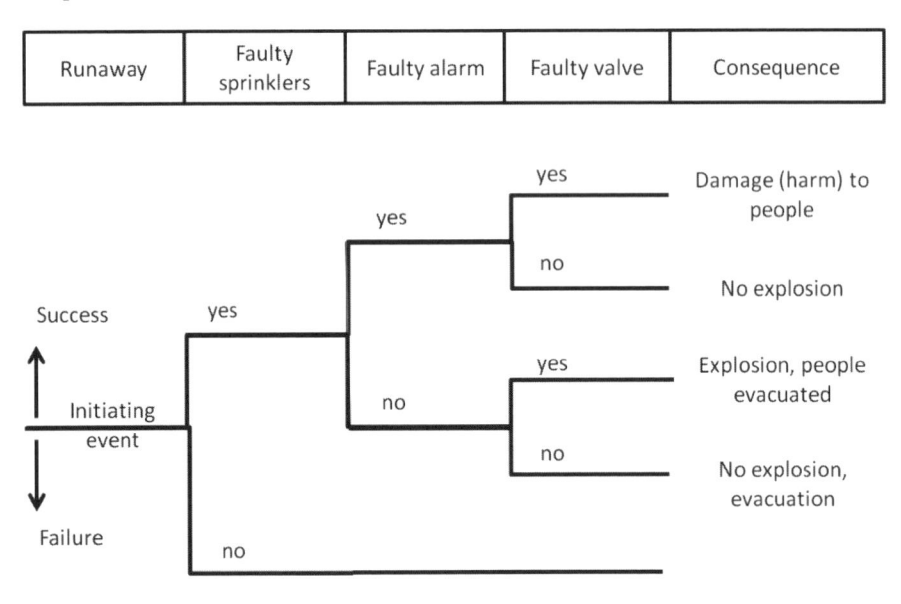

| Runaway | Faulty sprinklers | Faulty alarm | Faulty valve | Consequence |

Figure 5.10. *Consequence or event tree*

5.3.5. *Bow-tie diagram*

The bow-tie diagram is a model combining a fault tree with a consequence tree, centered on an undesirable event. It represents both the causes and the consequences of this event. The graphical representation of the fault tree is modified slightly to allow it to be oriented left to right (Figure 5.11). The undesirable event is known as the central event. The whole representation is centered on this event, implying that:

– certain events that may be consequences of the causes of the central undesirable event (CUE) will not be represented, if they themselves are not the cause of the CUE;

– the consequences represented on the right only depend on the CUE, and not on the causes that may have led to the occurrence of the CUE.

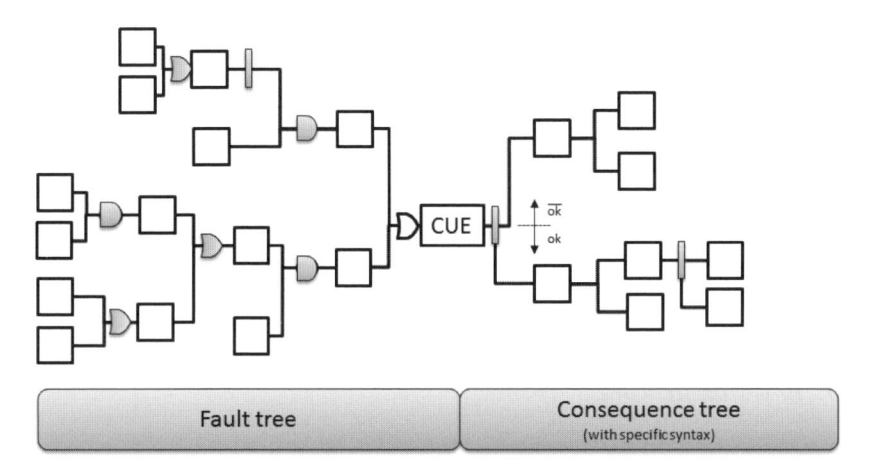

Figure 5.11. *Bow-tie diagram*

Barriers may be represented easily on this type of diagram by a vertical line through an arc, and this may be done in both the cause and consequence sections of the diagram.

The right-hand side of the diagram is not an event tree, in the strictest sense: usually, it is a tree representing all of the possible consequences of the CUE, and not a success–failure type event tree. A specific representation is used to model the effects of a barrier fault: in these cases, the arc is linked to the bottom of the barrier (see Chapter 15), and we obtain a representation with two alternatives, as in the case of a consequence tree.

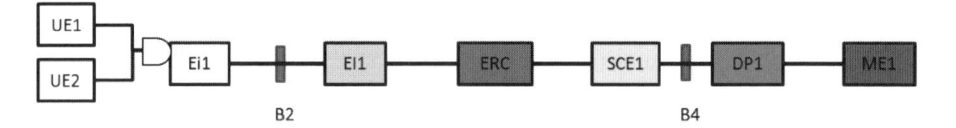

Figure 5.12. *Scenario extracted from the bow-tie diagram*

5.3.6. *Scenario*

A scenario is a causal chain of events, only using AND gates, moving from the earliest causes to the latest consequences. It constitutes a subset of the bow-tie diagram and takes the form of a possible pathway. Barriers may be represented by bars, as in the bow-tie diagram. This modeling approach is used in the layer of protection analysis (LOPA) method (section 14.6).

This model may be used to independently represent all of the possible faulty behaviors of a system, with one scenario per behavior, and to verify that one or more barriers exist in each case.

5.3.7. *Bayesian networks*

A causal event graph is a graph in which the nodes represent events and where oriented arcs describe causality relationships. A Bayesian network is an acyclic causal graph (cycles are not permitted). Each node is associated with a discrete random variable that may take a value from a finite set of values. A distribution function is associated with each variable. The arcs between two variables represent a probabilistic connection between two variables: the probability of a variable depends on that of its causes. This probabilistic link is described using a table of conditional probabilities, which contains the probability of different values of the variable associated with the node as a function of all of the possible input combinations. For example, if we consider the network shown in Figure 5.13, with the variable associated with A taking the values $\{low, medium, high\}$ and the variables associated with B and D taking values of $\{0, 1\}$, we obtain the following table for node D:

A	B	Pr(D = 1)
Low	0	0
Low	1	0.75
Medium	0	0.1
Medium	1	0.85
High	0	0.12
High	1	0.95

Using the properties of the network and conditional probability tables, we can calculate the probabilities of a value of a node as a function of the others, in the same way as for fault trees, but involving more complex relationships.

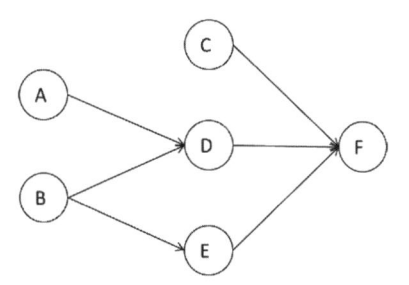

Figure 5.13. *Bayesian network*

NOTE 5.2.– It is possible to associate a real random variable with a continuous distribution function to each node, but in such cases analysis becomes considerably harder.

A Bayesian network may be seen as the generalization of a fault tree where the inferences of the nodes are determined by types of logic gates. Figure 5.14 represents the equivalent of an AND gate. The values of events A, B, C are of the type $\{0, 1\}$. The conditional probability table for node C to model the equivalent of the AND gate is as follows:

A	B	Pr(C = 1)
0	0	0
0	1	0
1	0	0
1	1	1

Bayesian networks give the possibility of more precise modeling. Let us consider an accident, described by a fault tree using an AND gate between two events, for example a complex task and poor training. The probability of this accident is $P(Accident) = P(A).P(B)$.

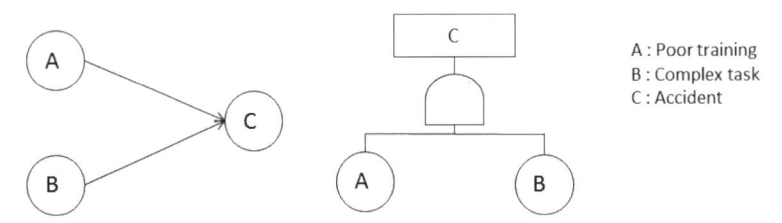

Figure 5.14. *Bayesian network and fault tree*

A : Poor training
B : Complex task
C : Accident

Using a Bayesian network, the probability of the accident is:

$$P(Accident) = P(Accident|A, B).P(A).P(B)$$

Using the table below, it is possible to show that there is a significant probability of an accident, even if the task is not complex, if poor training has taken place.

A: Complex task	B: Poor training	Pr(Accident|A,B)
0	0	0
0	1	0.3
1	0	0.05
1	1	1

Moreover, we can model the values of the variables A: complex task and B: poor training using more than two states, moving from $\{false, true\}$ to $\{low, medium, high\}$. We can then describe links using a three-entry table.

Bayesian networks therefore constitute a modeling type suited to the representation of complex connections. However, they can be difficult to construct, both from a structural perspective (links between variables) and in relation to conditional probability tables. They can only be used with suitable software tools and should be limited to use with specific problems.

5.4. Modeling dynamic aspects

5.4.1. *Markov model*

The models discussed above allow us to describe causal connections between the different events leading to the occurrence of a risk, but they do not allow us to take account of the order in which these events occur. This imposes limitations on the representation of fault-tolerant systems. Furthermore, they cannot be used to model the switch from a normal mode of operations to a faulty mode, or the return to the normal mode in case of repair.

EXAMPLE 5.4.– Let us consider a system made up of a main power supply, a backup power supply and a commutation system that connects the backup supply if the main supply fails. Faults in the commutation system are only important if they occur before the main power supply is disrupted; in this case, the backup supply would not work. The order in which faults occur is therefore significant, and this cannot be modeled using a fault tree.

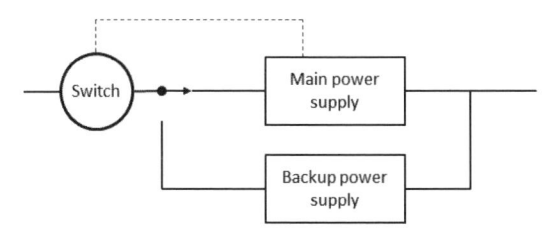

Figure 5.15. *System with backup power supply*

Moreover, when modeling the occurrence of a risk, it may be useful to model the state of different entities associated with the risk, particularly the switch from a normal to an abnormal state in which the damage- or loss-generating phenomenon occurs. This allows us to represent partially degraded states, which can be repaired.

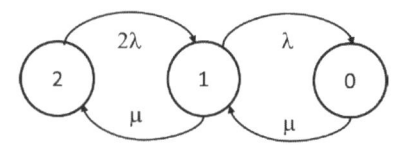

Figure 5.16. *Markov model*

A Markov model is a discrete state modeling represented by a graph. Each node represents a system state and each edge or arc represents a possible transition between two states. Each transition is associated with a coefficient representing the rate of transition between two states, generally the failure rate. Figure 5.16 represents the Markov model of a redundant system. The system is in state 2 when the two units are operating correctly, state 1 when one unit has failed and state 0 when both units have ceased to operate normally. The failure rate of each unit is noted λ. The repair rate is noted μ. Taking $P_0(t)$ as the probability that the system is in state 0, $P_1(t)$ the probability of state 1 and $P_2(t)$ the probability of state 2, the evolution of these probabilities is governed by the following system:

$$\frac{dP_2}{dt} = -2\lambda P_2 + P_1$$

$$\frac{dP_1}{dt} = 2\lambda P_2 - (\lambda + \mu)P_1$$

$$\frac{dP_0}{dt} = \lambda P_1 - \mu P_0$$

This model allows us to calculate the evolution of the probabilities of each state over time, and is used for specific parts of a system, where certain elements may be repaired, or where the order of failures is significant. Further details may be found in [ERI 05], for example.

EXAMPLE 5.5.– The Markov model of a power supply with a backup system is shown in Figure 5.17. The system starts out in state 0. It moves into state 1 if the main power supply fails, and then into state 2 if the backup fails. State 3 corresponds to the failure of the commutation system. This model allows us to take account of the order of failures. A fault in the main power supply following a switch failure places the system directly in state 2. It would also be possible to model repairs to the main power supply by adding an arc from 1 to 0 showing the repair rate.

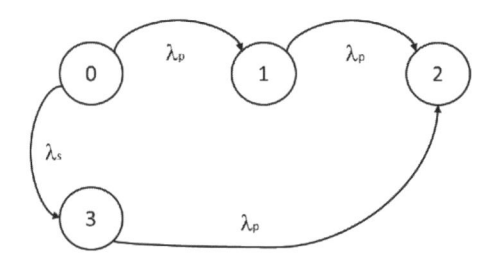

Figure 5.17. *Markov model of system with backup power supply*

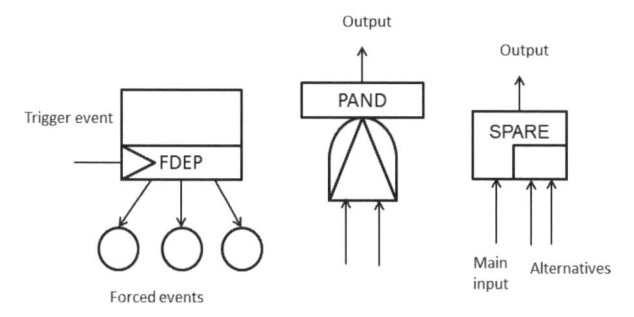

Figure 5.18. *Gates introduced by the dynamic fault tree*

5.4.2. *Dynamic fault tree*

Th dynamic fault tree (DFT) model was proposed in 1992 for use in analyzing fault-tolerant computer systems [DUG 92]. It allows for a more flexible and elaborate modeling of risk-tolerant systems by introducing the notion of dynamic gates. The summit event depends not only on the logical combination of basic elements, but also on the order in which these events become true. In practice, the DFT model allows us to effectively identify the minimum cut sets of events occurring in a certain order that will generate the summit event. The model introduces three types of gate:

– FDEP gate, which has a initiating event as input, the activation of which forces the output events. This gate has no output connected to the fault tree logic.

– PAND gate, a priority AND, for which the output is true if the inputs become true in a certain order.

– SPARE gate, for which the output becomes true if the main input and all of the secondary events are true and activated in the following order: principal, alternative 1, alternative 2, ..., alternative n.

EXAMPLE 5.6.– It is possible to model a system with a backup power supply using a DFT model. This model is shown in Figure 5.19.

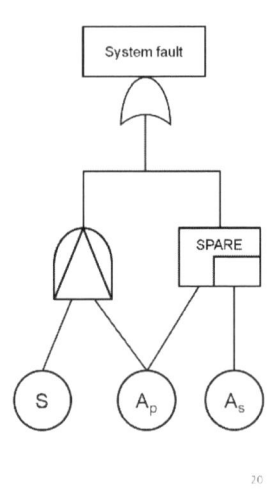

Figure 5.19. *Dynamic fault tree for the system with backup power supply*

5.5. Summary

In this chapter, we have presented the main methods used in modeling scenarios to describe a risk. This construction may be carried out directly, but in most cases it is integrated into a risk analysis approach after the risk identification stage. If the identification process is based on a sufficiently formal model, the scenario model may be deduced automatically using suitable software tools.

The causal tree method stands out from the other modeling methods discussed as it is used *a posteriori*, that is after an accident has taken place. Causal tree models are used to understand the conditions in which accidents occur.

Figure 5.20 presents a summary of the most widespread event graph models and the relationships between them. As we can see from the figure, these graphs are linked. They allow us to model a risk, identifying a CUE that represents the occurrence of the risk. Depending on the type of graph, the model may represent the causes, consequences, causes and consequences or a subset of causes or consequences of a risk. The event tree model is used to describe alternatives.

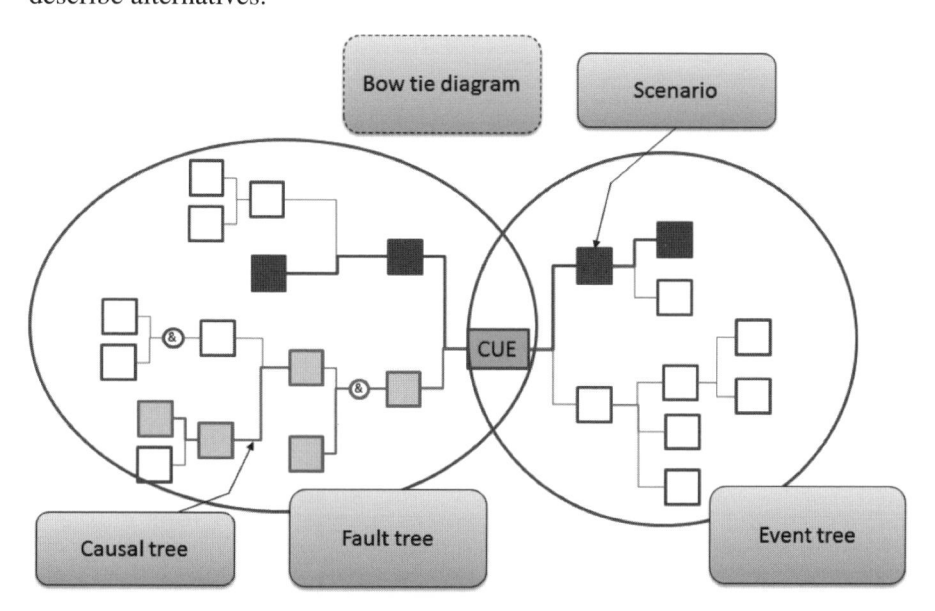

Figure 5.20. *Relations between the modeling approaches*

These event graph models are among the most widespread approaches. They allow us to formalize the description of a scenario, quantify risk levels and represent barriers.

Chapter 6

Measuring the Importance of a Risk

6.1. Introduction

The ISO31000 standard gives the following definition of risk:

"Risk is the effect of uncertainty on objectives, whether positive or negative, with regard to an expectation".

To assess the importance of a risk, we must therefore:

– evaluate the extent to which this event may occur, taking account of uncertainty, by defining the likelihood value of the effect linked to this risk;

– evaluate the importance of this effect, determining its impact, which is characterized by its severity in cases where the effect manifests itself in the form of losses.

The risk management process begins with a context establishment phase. At this point, we define levels of likelihood and severity, for which various options are available; these options will be discussed later in this chapter. We must also specify the method used to evaluate these levels and the type of measurement scale selected. This scale may be qualitative, semi-qualitative or quantitative depending on the type of risk in question and the data available.

Once this framework has been defined, each risk is assessed during the analysis phase, with a definition of its likelihood and severity. A risk model, which may show causes, consequences and/or barriers, may be used at this stage.

By combining the likelihood and the severity of a risk, either using a mathematical formula or a dual input table known as a risk matrix, we obtain a risk or criticality level. This level is then used in the risk evaluation phase to determine whether or not the risk is judged to be acceptable. In cases of unacceptable risk, the risk level is then used to determine what treatment measures should be envisaged.

Care must be taken in developing this rating approach. A variety of possibilities exist in terms of the following:

– The definition of the measurement scale, which varies according to:

- the nature of measurements, which may be qualitative or quantitative;

- the level of detail or "sharpness" of the scale;

- the way in which likelihood and severity are combined.

– The actual assessment of each risk, that is the way in which a risk is placed on the scale, both in terms of likelihood and gravity. This assessment may be carried out:

- in a broadly subjective manner, using the judgment of one or more experts, or through seeking consensus between the representatives of different interested parties, based on qualitative feedback;

- based on a more mathematical approach, using probabilistic calculation methods to calculate the likelihood of a risk based on its modeling and reliability data, and using physical models to calculate the intensity of the effects of a phenomenon and their severity, coupled with an effect-dose response curve.

The first type of approach is used for simple systems involving relatively simple scenarios. It is well suited to the evaluation of occupational risks, for example. The second approach allows a more rigorous approach suited to the analysis of complex systems, although, as we will see, the choice of different parameters and input data is a significant source of uncertainty.

In all cases, it is important to remember that the rating concerns the likelihood of damage, and not the likelihood of the event causing the damage. In cases where a single type of damage is possible and where there is no possibility of avoidance, the two values are clearly equivalent. In other cases, they are different, and the risk is generally overestimated.

To define risk quantification in a more precise manner, let us return to the definition given in Chapter 2 where a risk is defined by a modeling of a situation with uncertain consequences using a set of triplets with the form:

$$\{< S_i, P_i, C_i >\}_{i=1,..N}$$

where S_i designates a scenario, P_i its likelihood and C_i its consequences. The consequences of a scenario can vary widely and concern different targets. In a general manner, we can represent the consequences of a scenario by a set $C_i = \{c_{i1}, \ldots, c_{in_i}\}$ (Figure 6.1). For each consequence c_{iq}, we may define a severity value, or cost, and a likelihood $p_i(c_{iq})$. The likelihood of the scenario is determined by a set of n_i values.

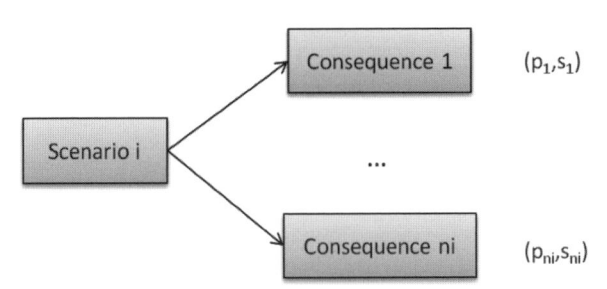

Figure 6.1. *Quantification of the risk described by a scenario*

In a large number of cases, only the most representative consequence of the scenario is identified in order to simplify the analysis process. In this case, the likelihood is evaluated using a single scalar value associated with the most representative consequence.

EXAMPLE 6.1.– Take the case of a person falling. The consequences may be defined by {bruising, reversible injury, irreversible injury and death}.

96   Risk Analysis

A probability may be defined for each of these consequences. In most cases, we would consider that the most representative consequence is reversible injury, and we evaluate the likelihood of this consequence.

The general principle used to measure a risk, based on an evaluation of its likelihood and the severity of damage, may be applied to all types of risk. However, its implementation differs considerably based on the nature of the risks and stakes involved. In the context of an analysis of occupational risks, for example, the most plausible damage is often deduced directly from the type of hazard in question, and likelihood is evaluated in a qualitative manner through consultation with the interested parties. In the case of a risk analysis concerning a more complex situation, however, it may be useful to construct scenarios in a more detailed manner in order to obtain quantified measurements of risk levels.

Examples of quantification for a number of major risks are shown in Figure 6.2.

Nature of catastrophe	Estimated probability (annual)	Maximum severity (number of deaths)
Earthquake	10^{-3} to 10^{-4}	10^6
Flood	10^{-2}	10^6
Tidal wave	10^{-2} to 10^{-3}	10^6
Tornado	10^{-2} to 10^{-3}	10^5
Dam rupture	10^{-3} to 10^{-4} (per site)	10^5
Chemical explosion and fire	10^{-5} to 10^{-7} (per site)	10^4
Nuclear reactor accident	10^{-7} (per site)	10^6
Mine accident	10^{-2} to 10^{-3}	10^2
Train accident	10^{-1} to 10^{-2}	10^2

Figure 6.2. *Examples of probability and severity values*

6.2. Assessing likelihood

6.2.1. *Presentation*

In general terms, the likelihood of a consequence of a scenario is evaluated either:

– by a symbolic value taken from a finite set of linguistic values, such as hihgly unlikely, unlikely, possible, common, or from a set of integer values $\{1, 2, ...N\}$,

– by a real numerical value x, $0 \leq x \leq 1$, representing a probability or degree of belief.

The first approach is used for systems where we have little precise information, or where a qualitative analysis is considered sufficient. This approach is, clearly, faster, but offers a reduced level of precision. The second approach is used for systems where quantitative feedback is available or for technical systems made up of components with available reliability data. It is suitable for using logical combinations of events which may be created during the analysis phase.

These approaches may also be combined to give a mixed approach, associating a probability interval with each linguistic value. The approach proposed in the May 2010 circular [MEE 10] on the analysis of industrial risks, discussed in section 6.6, falls into this category. Another example is proposed in standard MIL-882E [DOD 12] (Figure 6.4).

6.2.2. *Quantitative scale*

6.2.2.1. *Use of probability*

The notion of probability is relatively familiar, but also complex. The basic mathematical definition is given in Appendix 3. It may be interpreted in a number of ways, including:

– the axiomatic, or classic, approach;

– the frequentist, or objective, approach;

– the Bayesian, or subjective, approach.

The axiomatic approach is based on an analysis of the observed system. In the case of throwing a die, if we carry out a physical analysis of the die using precise measurement methods and we ensure that it is perfectly balanced, we can state that the chances of obtaining any given face of the die are strictly equal, and consequently their probability of occurrence is $1/6$. This approach is based on deductive reasoning; it does not require extensive experimentation, but the system in question must be suited to detailed analysis.

Using the frequentist approach, we suppose that the experiment is repeated a very large number of times, and we consider the relationship between observations characterizing a given event and the total number of experiments. This relationship is asymptotically equal to the probability of the event in question. To return to the example of the die, the relationship between the number of times a given face is obtained and the total number of throws gives us an asymptotic estimation of the probability of obtaining that face, the conceptual problem with this approach is that it is generally difficult to repeat an experiment a sufficient number of times in strictly identical conditions. It may be envisaged for simple technical objects, but is more difficult to apply in the case of complex systems or in situations where the damage created is significant.

The subjective, or Bayesian, approach considers the probability of an event as a subjective measure of the degree of belief in the possibility of occurrence of the event. In this case, it may be seen as a number representing the opinion of the analyst in relation to his or her existing knowledge. For example, based on his or her experience, an observer may give a probability that it will rain within the next hour.

In the context of risk analysis, it is rarely possible to repeat an experience a large number of times. In most cases, situations are complex, the conditions are unique and the event at the root of the risk is rare. Consequently, the subjective or Bayesian approach is generally used. Certain authors consider that subjectivity is strictly linked to the analyst: two different experts will give two different results depending on their preferences. Other authors consider that this subjectivity characterizes the lack of knowledge of the system: two experts with the same knowledge would give identical results [LIN 06].

In the first case, in addition to uncertainty linked to the system, subjective probability involves an aspect of uncertainty linked to the expert. The uncertain aspects linked to the system may be grouped into two categories [AVE 10]:

– random uncertainties, linked to the variability of the system;

– epistemic uncertainties, due to the lack of knowledge of the system.

In principle, it should be possible to reduce the second category of uncertainty by accumulating knowledge concerning the system; random uncertainties, on the other hand, are generally difficult to reduce. In certain cases, it is possible to do this, but in the context of risk analysis the systems concerned are complex. For this reason, it is often difficult to define a system and operating conditions for which we can accumulate the knowledge needed to better predict the behavior of the system.

Consequently, it is important to note that, in the context of risk analysis, while probability values may appear to be quantified and rigorous, a significant level of subjectivity remains. Results should, therefore, be treated with the same care used elsewhere in the risk management process. In these conditions, we also need to accept that we cannot systematically oppose calculated values and the subjective feelings of interested parties.

Efforts have been made to develop means of representing uncertainty concerning probability values themselves, using imprecise probabilities, approaches based on fuzzy logic or belief theory. These approaches are still at the research stage, but examples may be found in [HEL 04].

6.2.2.2. *Use of temporal frequency*

Using the frequentist definition of probability, we speak of frequency of occurrence, for instances, of the same experiment, repeated a large number of times. When considering initiating events or failures, this may be replaced by the notion of temporal frequency, defined as the number of occurrences of the event during a given period. This frequency is used to characterize the likelihood of an event, and cannot be used as a probability value without the addition of supplementary hypotheses (something which becomes evident if we consider that a frequency value may be greater than 1).

If, however, we have a frequency for the occurrence of an event, and we pose the hypothesis that the rate of occurrence of this event is constant, we can model the probability of occurrence of the event at time p by an exponential law:[1]

$$P(t) = 1 - exp(-F.t)$$

---

[1] See Appendix 4 for further details.

where F is the frequency of occurrence (number of times per unit of time) and t is the time of occurrence, expressed using the same units. If the event represents a system failure, then F is related to the mean time between failures (MTBF) by the relationship $F = 1/MTBF$. When the frequency is low and we calculate the probability by time unit ($t = 1$), the probability is equal to $1 - exp(-F) \approx F$.

A second aspect to consider is the concurrency of events. If the duration of an event is very short, there is little chance of two events taking place at the same time. In this case, the evaluation of the output of an OR gate with usual laws is a correct approximation. However, this is not the same for a AND gate, as the ouput event can only take place in the period in which the two input events overlap. If our calculations do not take account of this aspect, the probability of the output event will be strongly overestimated. In the case of two events, it is possible to approximate this probability using a simple relationship; for a number n of input events, however, we must use a dynamic fault tree [DEU 08].

6.2.3. *Qualitative scale*

In the case of a qualitative scale, the likelihood is evaluated using a finite number of values, generally between four and six, but as high as 10 for multi-risk scales such as those combining occupational and industrial risks. It is important to clearly define the specifications for each level in order to limit the effects of subjective interpretation by different analysts. Figure 6.3 shows a simple four-level scale. Figure 6.4 represents the scale used in the MIL-STD882E standard, which defines six levels.

Level	Semantics	Meaning
V1	Highly unlikely	This event seems extremely unlikely to occur. There are no known cases of occurrence.
V2	Unlikely	This event may occur, but is unlikely. An example of occurrence may be found, but for another installation.
V3	Possible	This event may occur. Cases of occurrence have been noted for this installation.
V4	Common	This often happens.

Figure 6.3. *Likelihood scale with four levels*

Using a symbolic scale, a minimal level is automatically included in the construction. When a risk is characterized by a probability of this level, it is not possible to represent a reduction in likelihood on the scale, and we remain at the minimum level. This does not occur when using a quantitative scale, where it is possible to represent an arbitrarily small probability value. To counteract this problem, we may add a level to the scale corresponding to an insignificant likelihood. This level is sometimes labeled IS (insignificant) or "eliminated", as in the table taken from standard MIL-STD-882. This level has a particular role to play in the risk matrix (Figure 6.12).

Description and level		Individual item	Group of items	Quantitative
Frequent	A	Occurs often in the lifetime of the item	Occurs almost continually	$10^{-1} \leq p \leq 1$
Probable	B	Occurs several times in the lifetime of the item	Frequently occurs	$10^{-2} \leq p \leq 10^{-1}$
Occasional	C	May sometimes happen in the lifetime of the item	Occurs several times	$10^{-3} \leq p \leq 10^{-2}$
Unlikely	D	Unlikely, but may occur in the lifetime of the item	Unlikely, but we may reasonably expect the event to occur in the lifetime of the item	$10^{-6} \leq p \leq 10^{-3}$
Very unlikely	E	Very unlikely – we may expect that the event will not occur in the lifetime of the item	Very unlikely, but not impossible	$p \leq 10^{-6}$
Eliminated	F	Cannot occur in the lifetime of the item. This category is used when potential dangers have been identified and subsequently eliminated.		

Figure 6.4. *Probability scale from standard MIL882E*

As we saw in the introduction to this section, it is also possible to use a mixed scale in which likelihood is evaluated in either a qualitative or numerical manner. In this case, we identify a probability interval corresponding to each level (Figure 6.4).

NOTE 6.1.– It is best to avoid odd value numbers if the values are not precisely defined; in these cases, if doubt is present, the analyst may be tempted to automatically select the median value.

6.2.4. *Determining likelihood values*

To determine a likelihood value, we may use:

– feedback, which is specific to the sector of activity or to a company;

– expert judgment [AYY 01];

– evaluation by consultation with interested parties.

In cases using quantitative values, we may use:

– databases or tables of reference (see Appendix 5);

– an aggregation of different probability values, based on the application of probability computation rules and on a causality model (fault tree, bow-tie diagram, etc.) built after a detailed risk analysis process.

These approaches will be discussed further in the remainder of this chapter for the specific case of occupational and industrial risks.

6.3. Assessment of severity

6.3.1. *Presentation*

To evaluate the importance of a risk, we need to evaluate the severity or the cost of the consequences. These consequences generally take the form of damage to targets with negative effects.

The first stage is to identify aspects requiring characterization, that is the type of target and the type of damage. The second stage consists of defining a measurement of this aspect. For risks associated with production systems, the targets and associated damages usually include:

– individuals: damage takes the form of an impact on health, up to death;

– populations: damage takes the form of an impact on the health of a group of individuals;

– the environment: damage to flora and fauna;

– quality of production;

– quantity of production;

– production time.

Before looking at the various indicators used, it should be noted that:

– when characterizing consequences, it is possible to measure the effective *quantity of damage* created by the scenario, or to measure the *intensity* of the phenomenon, without explicitly evaluating damage, which depends on the vulnerability of the target (Figure 6.5). In the case of an explosion, for example, we may evaluate the zone affected by overpressure at a given threshold, without explicitly evaluating the damage to populations;

– the notion of time is important, damage may occur immediately or a certain time after the event, as in the case of an illness detected some time after exposure to a toxic product or a failure linked to a previous shock;

– depending on the target and the damage, it may be possible to use a Boolean characterization (true/false) or a percentage value. For an individual, for example, damage taking the form of death may be true or false; for a population, however, we may talk of death rates.

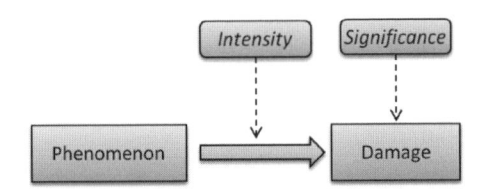

Figure 6.5. *Measuring consequences*

In a general manner, for a given risk, described by $R = \{< S_i, P_i, C_i >\}_{i=1,...,N}$, each scenario may have multiple consequences, noted $C_i = \{c_{i1}, ..., c_{in_i}\}$. An elementary consequence is noted c_{iq}. Each consequence corresponds to a type of damage which may be measured by a symbolic or numerical value, noted g_{iq}. The consequences of the scenario are characterized by a severity vector $[g_1, ..., g_{in_i}]$.

We may wish to aggregate different values in order to characterize a risk by a single value, and thus simplify the prioritization process. We might, for example, select the consequence with maximum severity. It is also possible to use the probability of each consequence c_{iq} of the scenario and combine it with g_{iq} to evaluate the risk. This approach is discussed further in section 6.4.

EXAMPLE 6.2.– Take the example of a dispersion of toxic gas. This has multiple consequences for man, flora and fauna. We may measure the irreversible effects for humans, deaths, the volume of water polluted and the cost for the company, for example. All of these consequences may be characterized by a severity value, and we may choose to retain the highest of these levels.

6.3.2. *Quantitative indicators*

The most widely used quantitative indicators are used to measure damage to human health. In the case of workplace accidents, we define the fatal accident rate (*FAR*):

$$FAR = \frac{\text{Number of deaths observed}}{\text{Number of hours of work}} 10^8$$

We also define the frequency rate:

$$FR_1 = \frac{\text{Accidents with leave} >1 \text{ day}}{\text{Number of hours worked}} 10^6$$

We may also take account of accidents which do not entail employee absence: FR_2. If we add incidents requiring treatment, the rate is known as FR_3.

For major industrial accidents, we define the number of persons affected at different intensity levels: irreversible injury, lethal effects and significant lethal effects (section 6.6.2).

It is also possible to measure damage to the environment, using the approach contained in the European accident scale (presented in the following section), for example, or to measure the financial cost of material damage to the company. Other quantitative indicators are presented in [JON 03].

6.3.3. *Qualitative indicators*

The severity of an accident may also be evaluated using a qualitative scale. Basic levels might include, for example, benign, serious, severe and catastrophic.

It is important to precisely define the meaning of each term for these scales to be usable. If the scale concerns damage to human health, the benign level might be defined as involving minor injuries, "serious" by "requires medical intervention" and so on. A scale developed for occupational risks is shown in section 6.5.

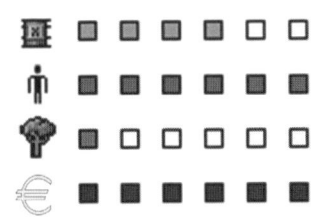

Figure 6.6. *Quantification using the European accident scale*

The MIL-STD-882E standard proposes a scale using four levels (Figure 6.7). Each of these levels is defined using a threshold in terms of human health, environmental impact and financial cost.

Description and level		Consequences of the undesirable event
Catastrophic	1	May produce one or more consequences of the type: - death, total permanent handicap, irreversible injury - significant environmental impact - financial loss greater than or equal to ten million dollars.
Critical	2	May produce one or more consequences of the type: - partial permanent handicap, work-related injury or illness leading to the hospitalization of at least three members of staff - significant, but reversible, environmental impact - financial loss greater than or equal to one million dollars but less than ten million dollars.
Marginal	3	May produce one or more consequences of the type: - work-related illness or accidental injury leading to the loss of at least one day of work - moderate reversible environmental impact - financial loss greater than or equal to 100K dollars but less than one million dollars.
Negligible	4	May produce one or more consequences of the type: - work-related illness or accidental injury leading to the loss of less than one day of work - minimal environmental impact - financial loss of less than 100 K dollars.

Figure 6.7. *Severity scale used in the MIL-STD-882 standard*

The European accident scale, available on the ARIA website, defines four categories of consequences:

– the quantity of hazardous matter released, measured in relation to the SEVESO threshold for the matter in question;

– human and social consequences, for damage ranging from minor oversights to death;

– environmental consequences, evaluating the impact on flora and fauna;

– economic impact, measured in euros.

Severity level	Human impact	Impact on means of production	Impact on quantity of production	Impact on quality of production
1 minor	No impact for staff	No influence on the process	No loss of production	No influence
2 major	Mission interrupted	Degraded operations	Loss which may be recovered by one team	Poor quality, but may be brought up to standard
3 critical	Unsafe situation	Influence on product quality or on safety	Loss which cannot be recovered by one team	Products cannot be brought up to standard
4 catastrophic	Risk of death or physical accident	Halt to production or accident	Loss from which no (or little) recovery is possible, or risk of accident	Halt in production or accidental creation of a dangerous product

Figure 6.8. *Severity scale used for damage of various*

For each category, an indicator is defined using a scale from 1 to 6. A total of 18 technical parameters are used to evaluate each of these indicators:

– Q1: quantity of substance released and Q2: quantity of substance involved in explosion;

– H1: number of deaths, H2: number of persons with serious injuries, H3: number of persons with minor injuries, H4: number rendered homeless, H5: number of persons evacuated, H6: number of persons left without drinking water, telephone, gas and electricity and H7: number of persons requiring sustained medical attention;

– E1: number of animals killed or to be put down, E2: proportion of rare species (animal or vegetable) destroyed, E3: volume of water polluted, E4: surface requiring cleaning and E5: length of waterfront requiring cleaning;

– C1: material damage to the buildings (or material assets), C2: loss of production for the factory production site, C3: damage and loss of production outside of the factory production site and C4: cost of cleanup operations.

For each of these parameters, six thresholds are defined, giving an index value of between 1 and 6 (Figure 6.23).

Other scales may be used (Figure 6.9). We will now present the scales used to evaluate occupational and industrial risks.

6.3.4. *Determining a severity value*

Let us assume that measurement scales have been defined for each category of consequences. For example, we might have a scale:

– Ec_1 for human consequences;

– Ec_2 for environmental consequences, etc.

Our aim is now to characterize the consequences for each case. Note that, in numerous cases, a single scale is used, for example the human consequence scale for industrial risks.

To obtain a severity value for the consequences of occurrence of a scenario related to a risk defined by $< S_i, P_i, C_i >$ with $C_i = \{c_{i1}, \ldots, c_{in_i}\}$, we take the following approach for each c_i:

– for each category of consequences, identify the type of damage, for example injuries and death, value of material damage;

– in cases where several damage types correspond to the category, break down the category into several different consequences, each corresponding to a type of damage. For example, if the consequence is $c_{iq} = \{\text{injuries}\}$, and the scale includes "irreversible injury" and "benign injury" categories, we need to re-express the consequences: $c_{iq} = \{\text{irreversible injury}\}$ and $c'_{iq} = \{\text{benign injury}\}$;

– place the consequence on the measurement scale. There are two possible cases: either the damage is Boolean in nature, in which case we simply need to know if it exists in order to position the consequence, or the damage is quantified by a number, in which case we must determine its importance in order to place the severity of the consequence.

Level	Risk levels
1	Catastrophic
2	Major
3	Moderate
4	Minor
5	Insignificant
	Occupational health
5	Death
4	Permanent injury
3	Hospitalization required
2	Nurse's office
1	No treatment –recorded in incident log
	Company reputation
3	Long term damage to company image
2	Negative publicity on a national scale (short to medium term)
1	Questions raised in media
	Environment
4	Catastrophic irreversible impact on the natural environment and on infrastructure, requiring long-term repair work
3	Significant irreversible impact on the environment and on infrastructure requiring considerable repair work
2	High reversible impact on the environment and on infrastructure
1	Low reversible impact on the environment and on infrastructure
	Financial
3	Very high financial losses, program or operation may be unable to continue
2	Significant financial losses, harming short or medium term profitability of a program or operation
1	Minor financial losses
	Judicial aspects
3	Major litigation leading to criminal prosecution
2	Major litigation for sizeable infractions
1	Minor litigation for minor legal or regulatory infractions

Figure 6.9. *Severity scales*

EXAMPLE 6.3.– In the case of occupational risks, we might use the following set of damage values: {discomfort, benign injury, irreversible injury, death}, each of which is associated with a severity level varying from 1 to 4. Considering a risk where the possible consequence is death, the severity level is 4.

EXAMPLE 6.4.– For an industrial risk, we might use the following set of damage values: {irreversible effects, lethal effects, strong lethal effects}. Taking the example of a risk of explosion, the possible consequences might be of two types: c_1 = irreversible effects and c_2 = lethal effects. For each consequence, we evaluate the number of potential victims using a model that allows us to evaluate the overpressure caused by the explosion and a map showing the distribution of the population. The qualitative severity level is determined based on the number of victims using a table provided in relevant regulations.

6.4. Risk assessment

A risk is assessed by combining the likelihood and severity values. There are two possible approaches. The first approach is quantitative, which consists of defining the criticality of the risk. The second approach is qualitative and involves the use of a risk matrix. The two approaches may be combined to give a mixed approach, on the condition that certain precautions are taken.

6.4.1. *Criticality*

Criticality is defined as the mathematical expectation of losses (or gains), which are measured by the severity of a risk. If a risk is defined by:

$$R = \{< S_i, P_i, C_i >\}_{i=1,\ldots,N}$$

and the consequences are expressed as $C_i = \{c_{i1}, \ldots, c_{in_i}\}$, the criticality of a scenario s_i is expressed as follows:

$$crit(s_i) = \sum_{q=1}^{n_i} p(c_{iq}) \cdot g(c_{iq})$$

If the risk is characterized by a single scenario and a single consequence type, the expression is simplified, and we return to the expression given in section 2.10:

$$crit(R) = p.g$$

Different risks may be represented on a probability-severity graph (Figure 6.13) to obtain a risk mapping.

6.4.2. *Risk matrices*

When a qualitative risk assessment method is used, criticality may be evaluated using a dual entry table known as a risk matrix. Another approach consists of assigning a numerical value to each qualitative value and multiplying the two numerical values.

Risk analysis is often carried out using likelihood levels numbered from 1 to n and probability levels labeled in a similar way. The criticality, or risk index, is evaluated by multiplying the two indices. This approach should be used with care, as the scales involved may not be coherent. For a risk of level ($p = 1$, $s = 4$) and another of level ($p = 4$, $s = 1$), we obtain the same score 4, signifying that the two risks have the same importance. If we take account of the meaning of the symbols, this is not necessarily the case. Consequently, it is often better to use a risk matrix, which allows a certain degree of flexibility, and not to accord the same level of two risks appearing in squares which are symmetrical in relation to the diagonal, i.e. avoid giving the same importance to risks which have the same qualitative criticality but different values for p and s. An example is shown in the matrix in Figure 6.10.

	V1 Highly unlikely	V2 Unlikely	V3 Possible	V4 Common
G4 Catastrophic	Medium	High	Very high	Very high
G3 Critical	Medium	High	High	Very high
G2 Major	Low	Medium	High	High
G1 Minor	Low	Low	Medium	High

Figure 6.10. *Risk matrix*

It is often preferable to use the risk matrix directly without calculating the product of indices. Figure 6.11 shows a risk matrix including an IS level. The IS probability column leads to an insignificant risk level, whatever the level of severity involved. The same applies to the IS severity level. This approach is also used in the matrix proposed in the MIL-STD-882E standard (Figure 6.12) for likelihood levels.

	IS Insignificant	V1 Highly unlikely	V2 Unlikely	V3 Possible	V4 Common
G4 Catastrophic	IS	Medium	High	Very high	Very high
G3 Critical	IS	Medium	High	High	Very high
G2 Major	IS	Low	Medium	High	High
G1 Minor	IS	Low	Low	Medium	High
IS Insignificant	IS	IS	IS	IS	IS

Figure 6.11. *Qualitative matrix*

	Catastrophic (1)	Critical (2)	Marginal (3)	Negligable (4)
Frequent (A)	High	High	Serious	Medium
Probable (B)	High	High	Serious	Medium
Occasional (C)	High	Serious	Medium	Low
Unlikely (D)	Serious	Medium	Medium	Low
Improbable (E)	Medium	Medium	Medium	Low
Eliminated (F)	Eliminated				

Figure 6.12. *Risk matrix from MIL-STD-882E standard*

NOTE 6.2.– Several grids may be defined as a part of the same analytical process, depending on the types of targets or consequences involved.

6.4.2.1. *Risk matrices and quantitative risk measurements*

Risk matrices define levels using qualitative values and do not always allow us to define risk levels which are consistent with the criticality value, determined using quantitative values, in order to make optimal decisions [COX 08]. To maintain consistency, a certain number of properties must be verified:

– It should not be possible to increase the risk level by more than one symbolic level by an infinitesimal increase in probability or severity (property of weak consistency); consequently, two squares with two shared angles cannot have more than one level of difference;

– The lowest line or column of severity or probability should be labelled with the lowest risk level, as the whole line or column represents risks with the same criticality.

Even in cases where consistency is verified, a risk matrix does not always lead to an optimal distribution of effort. This effect is due to the discretization of continuous values, which creates a threshold effect and allows a single square to represent risks of differing criticality.

However, these properties assume that it is possible to determine an objective quantified measurement for each case. In many cases, we only possess qualitative measurements, and our reasoning cannot be based on the idea that a quantitative criticality measurement exists to characterize each risk, permitting precise prioritization or optimization of reduction efforts.

Finally, note that some of the matrices presented below do not verify these properties and are therefore only suitable for qualitative analysis or decision-making.

6.4.3. *Acceptability of a risk*

In cases of quantitative rating, the criticality value is used to determine whether or not a risk is acceptable in relation to predefined thresholds. Isocriticality curves define the Farmer criterion (Figure 6.13). Above the curve, which defines a risk level $p.s = limit$, the risk is considered to be unacceptable; anything below the curve is considered to be acceptable.

It is also possible to use the ALARP principle presented in section 2.17 to evaluate whether or not a risk is tolerable.

Measuring the Importance of a Risk   113

In the case of a qualitative evaluation, the risk matrix is used to define acceptable and unacceptable levels of risk. The level of acceptability is defined based on the qualitative risk level (Figure 6.14). This matrix is important and must be discussed with all stakeholders.

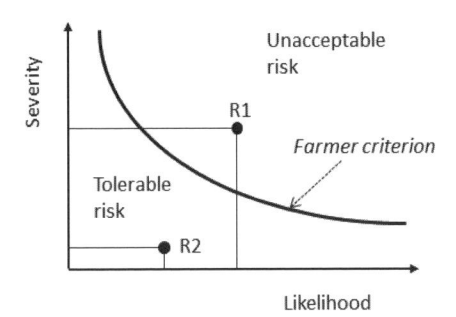

Figure 6.13. *Farmer criterion*

	IS Insiginifcant	V1 Highly unlikely	V2 Unlikely	V3 Possible	V4 Common
G4 Catastrophic	IS	Acceptable Reduce if possible	Acceptable Reduce rapidly	Unacceptable	Unacceptable
G3 Critical	IS	Acceptable Reduce if possible	Acceptable Reduce rapidly	Acceptable Reduce rapidly	Unacceptable
G2 Major	IS	Acceptable	Acceptable Reduce if possible	Acceptable Reduce rapidly	Acceptable Reduce rapidly
G1 Minor	IS	Acceptable	Acceptable	Acceptable Reduce if possible	Acceptable Reduce rapidly
IS Insignificant	IS	IS	IS	IS	IS

Figure 6.14. *Acceptability matrix*

6.5. Application to the case of occupational risks

6.5.1. *Probability assessment*

The assessment system used for occupational risks must remain relatively simple. A qualitative approach is therefore appropriate. Subjective qualitative probability may be used to express the judgment of the analyst and of stakeholders. A scale using qualitative severity levels allows us to represent the importance of different damaging effects, and also enables the integration of feedback.

To obtain clear rating results, the risk must be well-defined with clearly specified consequences. If we speak of a risk of falling, for example, with no further details, the use of a single (p, g) pairing for rating may be difficult. We need to give further details, for example "risk of falling leading to fracture and/or bruising", then rate the risk for each of the possible consequences.

Finally, the consequences of a risk may only become apparent after a certain time, and not immediately after the occurrence of the undesirable event, which may itself last for a certain period of time in the case of a chronic risk. This aspect must be taken into consideration.

The evaluation of occupational risks begins with an identification of these risks, which are then evaluated in order to prioritize them and to identify those which are unacceptable. As stated above, a qualitative scale is entirely appropriate in these cases.

A four- or five-level rating system is often used. It is sometimes preferable to choose an even number of levels to prevent the analyst from choosing the median level by default.

Figure 6.15 shows a proposed scale in which levels are defined by code letters and semantics. The proposed code takes a non-numerical form in order to remove the possibility of using level numbers directly in criticality calculations. The "meaning" aspect is used to give a description of the likelihood of the events associated with each probability level.

Level	Semantics	Meaning
A	Common	More than 3 times/month
B	Likely	From approx. 3 times/year to 3 times/month
C	Unlikely	From approx. once every 3 years to 3 times/year
D	Very unlikely	From approx. once every 30 years to every 3 ans
E	Extremely (or highly) unlikely	Approx. once every 30 years (installation lifetime)
IS	Insignificant	Conditions of occurrence eliminated, or occurrence practically impossible

Figure 6.15. *Definition of likelihood levels*

It is important to remember that this probability applies to the occurrence of damaging consequences, and not to the phenomenon or exposure to the phenomenon. In the case of diseases which develop after a certain period, for example, we evaluate the probability that the disease will develop if the employee has been exposed to the risk. This probability must not be confused with the probability or frequency of exposure, although the values are linked. In the same way, in the case of an accident leading to lower back pain due to carrying of excessive loads, we consider the possibility of developing this back problem, given the activity of the operator and not the frequency of this activity.

In cases with several possible consequences, for example electric shock or electrocution following work on connected electrical material, we specify the damage type for consideration, if one of these types is considerably more usual. In other cases, we must characterize the probability of each damage type separately.

In certain cases, we can break down the evaluation of the probability of damage occurring by evaluating the frequency of exposure and the probability that damage will occur per unit of exposure. This probability is also known as the risk control level. A dual entry table may then be constructed to determine the probability (Figure 6.16).

		Probability of occurrence of damage during the activity Or Risk control level		
		Rare Risk occurs very rarely Ex: fall on clean, dry floor surface or High risk control level	Frequent Risk often occurs Ex: fall when working on a poorly installed ladder or Medium risk control level	Very frequent Systematic occurrence of risk Ex: exposure to toxic element or Low risk control level
Frequency of activity	Very low, a few minutes per day, or once a week	E	D	C
	Low, 10 % of the time, a few times a week	D	C	B
	High, 10 % to 90 % of the time, several times a week	C	B	A
	Very high, more than 90 % of the time	B	A	A

Figure 6.16. *Indirect evaluation of probability*

By convention, the probability value is increased when the risk concerns a situation in which regulations are not respected. If an electrical cabinet is left

116   Risk Analysis

open in a workshop, for example, the risk of electric shock may be fixed at a high value, even if this does not strictly reflect the reality. The purpose of this convention is to show that these risks are unacceptable.

EXAMPLE 6.5.– Consider the example of two risk-generating situations in a semi-industrial bakery. Take the risk of falling from a standing position, leading to spraining, with a large quantity of flour dust on the floor; from experience, we know that this situation occurs frequently, and we give it a rating of B. Concerning the risk of an allergic disease related to flour, after consulting a specialist who indicates that, after around 10 years, all employees are obliged to find work in another domain of activity, we rate the risk of developing an allergy in the following 10 years at level A.

6.5.2. *Severity assessment*

The severity of occupational risks is assessed by considering effects on human health. There are two possible cases: work-related accidents or work-related illness. Injury in professional trips may be rated in the same way as workplace injuries.

The damage levels for accidents may be defined based on the type of treatment, the duration of leave or the level of invalidity they entail. For certain illnesses which present a long time after exposure, only the level of invalidity may be considered.

The severity of a risk depends not only on the level of damage, but also on the number of persons concerned by the level of damage; for example, we might choose to increase the severity level as this number increases. Figure 6.17 shows a possible scale.

6.5.3. *Risk matrices*

Figure 6.18 represents a matrix that can be used to define the acceptability of a risk. This matrix allows us to define a strategy for each severity/probability pairing. Five levels are used:

– A1: unacceptable;

– A2: reduce as a priority;

– A3: reduce if possible;

– A4: tolerable;

– IS: insignificant risk.

Level	Damage	Number of persons		
		1	2 to 10	>10
L1	Death, permanent total invalidity	S1	S1	S1
L2	Level of permanent invalidity > 10 %, sick leave > 30 days, permanent restrictions, retraining for new occupation	S2	S1	S1
L3	Permanent invalidity < 10 %, sick leave from 8 to 30 days, temporary restrictions	S3	S2	S1
L4	Sick leave < 8 days, Accident declared to social security without need for sick leave, effects reversed after treatment and healing period	S4	S3	S2
L5	Accident declared to register of benign accidents Discomfort or fatigue	S5	S5	S4

Figure 6.17. *Severity scale with consideration given to the number of targets*

	NS	E	D	C	B	A
S1	NS	A2	A1	A1	A1	A1
S2	NS	A3	A2	A1	A1	A1
S3	NS	A3	A3	A2	A1	A1
S4	NS	A4	A3	A3	A2	A1
S5	NS	A4	A4	A4	A3	A2

Figure 6.18. *Acceptability matrix*

The last level (IS) level is used when the risk is insignificant or when the hazard source has been eliminated. This would be the case if, for example, a toxic solvent used in a technical operation was replaced by a non-toxic product. Level A4 characterizes risks where no action needs to be taken, due to the low severity and low likelihood of the risk; in these cases, it is not worthwhile to devote company resources to reduction activities. Level A1, on the other hand, is used for risks which are clearly unacceptable: these risks must be eliminated before the activity can be allowed to continue. The intermediary levels, A2 and

A3, are used to prioritize risks which should be reduced, where possible, with priority given to A2 category risks. The proposed matrix can, obviously, be adapted to suit different specific situations.

6.6. Application to the case of industrial risks

Damage caused by major industrial risks is generated by three types of physical phenomena: fire, explosions and the dispersion of toxic products.

The French regulations on risk prevention published on May 10, 2010 [MEE 10] propose a semi-qualitative approach to measuring probability and severity. Severity is evaluated by considering a combination of two aspects, intensity and vulnerability. The principle used in this approach may be extended for use with other types of risks.

6.6.1. *Probability assessment*

Probability is determined using a mixed qualitative and quantitative approach. The scale includes five linguistic levels, each of which corresponds to an interval of numerical probability values. For simple installations, a qualitative approach is used, and the level is determined using the explanatory text given in the table: for example, if the event has already occurred during the lifetime of the installation, then value B is selected.

For more complex systems, a detailed model of the occurrence of the hazardous phenomenon is constructed during the risk analysis phase. This model may be represented by a bow-tie diagram, for example (Chapter 13). This model can then be used to obtain a numerical evaluation of the probability. The probability value given in the table is calculated for a time period of 1 year.

6.6.2. *Severity assessment*

The severity of a risk is calculated based on the intensity of the phenomenon and its consequences for the population:

– taking account of the type of phenomenon and using a suitable mathematical model, we identify the geographical area affected by the phenomenon with a certain level of intensity. There are three of these intensity levels, as shown in Figure 6.20;

– for each area, we determine the number of inhabitants who may be concerned, using geographical data and a certain number of rules determined in the May 10, 2010 regulations;

– we then use Figure 6.21 to determine the severity class of the scenario based on the number of persons exposed. The severity is determined for each type of phenomenon, and we retain the maximum level.

Assessment type	Probability class				
	E	D	C	B	A
Qualitative (the cited definitions only apply if the number of installations and quantity of feedback are sufficient)	Event is "possible, but extremely unlikely": Not impossible in the light of current knowledge, but has not occurred anywhere in the world over a long period of installation use	Event is "highly unlikely": Has already occurred in this sector of activity, but corrective measures have been taken which significantly reduce the probability	Event is "unlikely": A similar event has already occurred in the sector of activity or in this type of organization at global level. Possible corrections made in the intervening period do not guarantee a significant reduction in probability	Event is "likely": Has happened and/or may happen in the lifetime of the installation	Event is "common": Has happened on the site in question and/or may happen several times during installation lifetime, in spite of corrective measures which may have been taken
Semi-quantitative	This scale takes an intermediate place between the qualitative and quantitative and allows mastery measures to be taken into consideration.				
Quantitative (per unit and per year)	10^{-5}	10^{-4}	10^{-3}	10^{-2}	10^{-1}

Figure 6.19. *Probability scale for industrial risks*

	Significant Lethal Effect Thresholds (SLET)	Lethal Effect Thresholds (LET)	Irreversible Effect Thresholds (IET)	Indirect effect thresholds
Thermic	8 KW/m^2 1800 [(KW/m^2)$^{4/3}$]sec	5 KW/m^2 1000 [(KW/m^2)$^{4/3}$]sec	3 KW/m^2 600 [(KW/m^2)$^{4/3}$]sec	-
Toxic (exposure: 1-60 min)	CL 5%	CL 1%	IET	-
Overpressure	200 mbar	140 mbar	50 mbar	20 mbar

Figure 6.20. *Intensity thresholds according to types of effect*

120 Risk Analysis

Severity scale

Severity scale	Zone delimited by SLET	Zone delimited by LET	Zone delimited by IET
Disastrous	> 10 persons exposed	> 100 persons exposed	> 1000 persons exposed
Catastrophic	1-10 persons exposed	10-100 persons exposed	100-1000 persons exposed
Important	Max. 1 person exposed	1-10 persons exposed	10-100 persons exposed
Serious	0 persons exposed	Max. 1 person exposed	1-10 persons exposed
Moderate	No zone of lethality outside of the establishment		< 1 person exposed

Figure 6.21. *Severity scale*

6.6.3. *Risk matrices*

The risk matrix used for industrial risks is known as the MMR grid. It is defined in the 2010 regulations in relation to the assessment criteria used for the risk mastery approach in the case of accidents with the potential to occur in "SEVESO" establishments. Depending on the (p,s) pairs of different scenarios and the associated risk level (green, MMR or NO), different risk treatment measures must be implemented. The mechanism is the same as that which is habitually used in this context, but with certain particular features:

– the cell (Disastrous, E) is used in a different manner depending on whether or not the installation already exists;

– if at least five scenarios are found in MMR2 level squares, this is considered as a level 1 NO.

	E	D	C	B	A
Disastrous	Partial NO (new sites) MMR 2 (existing sites)	NO 1	NO 2	NO 3	NO 4
Catastrophic	MMR 1	MMR 2	NO 1	NO 3	NO 4
Important	MMR 1	MMR 1	MMR 2	NO 1	NO 2
Serious			MMR 1	MMR 2	NO 1
Moderate					MMR 1

Figure 6.22. *Risk matrix for industrial risks (MMR grid)*

Measuring the Importance of a Risk   121

	Quantities of dangerous substances	1	2	3	4	5	6
Q1	Quantity Q of substance lost or released in relation to the "Seveso" threshold (%)	Less than 0.1	0.1 to 1	1 to 10	10 to 100	1 to 10 times the limit	More than 10 times the limit
Q2	Quantity of substance having participated in the explosion (equivalent in TNT) (t)	Less than 0.1	0.1 to 1	1 to 5	5 to 50	50 to 500	Over 500
Human and social consequences		**1**	**2**	**3**	**4**	**5**	**6**
H3	Total number of deaths	-	1	2 to 5	6 to 19	20 to 49	50 and over
H4	Total number injured with hospitalization >24h	1	2 to 5	6 to19	20 to 49	50 to 199	200 and over
H5	Total number injured with hospitalization <24h	1 to 5	6 to 19	20 to 49	50 to 199	200 to 999	1000 and over
H6	Total number homeless or unable to work	-	1 to 5	6 to 19	20 to 99	100 to 499	Over 500
H7	Number of residents evacuated or confined to their homes	-	Fewer than 500	500 to 5000	5000 to 50000	50000 to 500000	Over 500000
H8	Number of persons without drinking water, electricity, gas, public transport for over 2h x number of hours	-	Fewer than 1000	1000 to 10000	10000 to 100000	100000 to 1000000	Over 1000000
H9	Number of persons having undergone extended medical supervision	-	Fewer than 10	10 to 50	50 to 200	200 to 1000	Over1000
Environmental consequences		**1**	**2**	**3**	**4**	**5**	**6**
Env1	Quantity of wild animals killed, injured or rendered unfit for human consumption (t)	Under 0.1	0.1 to 1	1 to 10	10 to 50	50 to 200	Over 200
Env 2	Proportion of animal or vegetal species destroyed (%)	Under 0.1	0.1 to 0.5	0.5 to 2	2 to 10	10 to 50	Over 50
Env 3	Volume of water polluted (in m3)	Under 1000	1000 to 10000	10000 to 100000	100000 to 1 million	1 million to 10 million	Over 10 million
Env 4	Surface area of soil or underground water surface requiring cleaning (in ha)	0.1 to 0.5	0.5 to 2	2 to 10	10 to 50	50 to 200	Over 200
Env 5	Length of water front or water channel requiring cleaning (in km)	0.1 to 0.5	0.5 to 2	2 to 10	10 to 50	50 to 200	Over 200

Figure 6.23. *European accident scale*

	Economic consequences	1	2	3	4	5	6
€15	Property damage in the establishment (in millions of euros)	0.1 to 0.5	0.5 to 2	2 to 10	10 to 50	50 to 200	Over 200
€16	Loss of production within the establishment (in millions of euros)	0.1 to 0.5	0.5 to 2	2 to 10	10 to 50	50 to 200	Over 200
€17	Property damage or loss of production outside the establishment (in millions of euros)	–	0.05 to 0.1	0.1 to 0.5	0.5 to 2	2 to 10	Over 10
€18	Cost of cleaning or environmental rehabilitation measures (in millions of euros)	0.01 to 0.05	0.05 to 0.2	0.2 to 1	1 to 5	5 to 20	Over 20

Figure 6.23. *(Continued) European accident scale*

Chapter 7

Modeling of Systems for Risk Analysis

7.1. Introduction

7.1.1. *Why model a system?*

When carrying out risk identification, analysis, assessment and treatment, we need to be able to describe a certain number of elements:

– the event or events that characterize the risk;

– the sources and targets of the risk;

– causes and consequences;

– likelihood and severity.

All of these aspects may be described using the methods described in Chapter 5. The model obtained in this way is known as a risk model or dysfunction model. It allows us to characterize the risk by describing the causality links between potential events creating undesirable consequences.

When analyzing a risk, it is also useful to model the system being analyzed, the object of the study. This modeling is used to construct a shared and formalized representation of the system. It improves understanding of the system, promotes communication between the different parties involved and allows us to structure the risk analysis process. The main aspects to describe are:

– the functions of the system, their interconnections and the specification of performances and ranges of operations;

– the structure of the system, identifying its components, whether material, human, organizational or information;

– the relationships between these elements;

– the behavior of the elements, used to describe how the whole system operates.

Moreover, this modeling must be sufficiently flexible to allow:

– an iterative approach: the system model often needs to be modified in the course of the risk analysis process;

– the choice of a type of approach and the level of detail used in modeling; it should allow us, for example, to choose a systemic or analytical approach.

7.1.2. *The modeling process*

A model of a system is a representation of reality (Figure 7.1). An infinite number of models are possible, depending on the perspective from which we approach the task. The model depends on the selected modeling approach. This approach is described by a metamodel, that is a model of a model. Depending on our objectives, the same reality may be represented in very different ways.

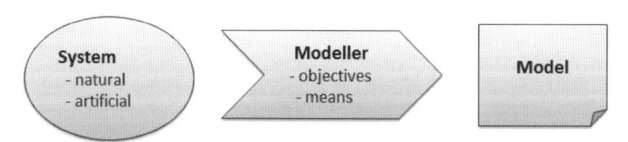

Figure 7.1. *The modeling process*

A building, for example, might be described:

– by a set of pictures for a decorator;

– by a set of plans (building plan, electrical circuit plans, etc.) for a construction engineer;

– by a mathematical model used to calculate the evolution of temperature for a heating engineer;

– by a number of potential occupants and a localization for an emergency service operative.

In the same way, very different representations of an industrial installation may be used. Take the example of an industrial fermentation process. The modeling of this process may express very different perspectives:

– For a biochemist, the process is represented in terms of chemical reactions.

– A process engineer might see the process in terms of transfers of matter and energy.

– The operator sees the installation as their daily work environment, and his or her skills are partly coded by others.

– The quality control manager essentially seeks to define a list of key variables to use in quality control.

– An automation engineer sees the process in terms of dynamics and systemics, that is as a set of interconnected systems evolving over time.

– An environmental and safety specialist might see the system in terms of possible deviations, probabilities of occurrence and means of preventing and limiting deviations in the process.

– An ergonomist might consider the user interface to be an aspect that can be crucial in risk prevention.

– The operations manager sees the process in terms of planning, production capacity, workers, material supplies, etc.

– A maintenance operative might consider the system in terms of materials: tanks, pipes, valves, automata, etc., which require preventative or curative intervention.

Each of the parties listed above is concerned with a different aspect of the system. If any of these individuals creates a model, taking account of his or her objectives and abilities, the representation produced will be specific to its creator. In the context of risk analysis, we must give consideration to the interactions between these approaches, and represent aspects taken from these different viewpoints. A valve problem, for example, resulting from a maintenance problem, may create a deviation in the procedure, which, given the energy transfer involved, will cause an automatic system to deviate from its zone of operations; this will require intervention by the operator in order to correct it, and will lead to delays in production and a loss of quality.

As we are not interested in constructing a detailed model describing all of these view points (for reasons of complexity and time), the selected modeling approaches must be suited to allow a simple and common representation that will act as a basis for risk analysis.

The approaches generally used in the context of risk analysis include:

– systemic modeling: the system requiring analysis is divided into subsystems, each of which is analyzed as a whole without being described explicitly;

– functional modeling: we aim to identify the different functions of the system, allowing us to examine the causes and effects of potential dysfunctions, without requiring detailed knowledge of all of the components;

– structural modeling: a list of physical components, materials, information and organizational elements and human actors is created in order to examine the causes and effects of potential faults and to identify the hazard sources represented by these elements;

– behavioral modeling: a model of the dynamic behavior of the system, that is its evolution over time, is created in order to study abnormal behaviors. This approach requires considerable effort in the development of the model and is only used in certain specific cases.

These approaches may be used in succession, giving an increasingly detailed representation of the system. In the remainder of this chapter, we will consider these modeling approaches and look at the way in which the obtained models interact with the dysfunction model.

7.2. Systemic or process modeling

7.2.1. *Principle*

Up to this point, we have used the term "system" to designate the object of analysis or the installation requiring analysis. In the context of systemic modeling, the notion of the system has a precisely defined meaning.

DEFINITION 7.1.– *A system [LEM 94] (from the Greek* sustëma, *which means assembly) is an organized whole, made up of critically linked elements that can only be defined in relation to each other, as a function of their place within the whole. The whole is greater than the sum of its parts, or, in other words, the properties of the complex system are more than the sum of the properties of its components. This is known as the emergence phenomenon. In other words, a system is a complex and structured set of mechanical, electronic, information or other components, which are in permanent interaction and which fulfill one or more functions.*

The idea of systemic modeling was developed by Eriksson [ERI 97], an approach that is discussed in Appendix 6. It is also at the center of the process modeling approach proposed in the ISO9001 standard.

This modeling approach allows us to describe a complex installation by dividing it into subsystems or processes, without needing to describe these parts in detail. It enables synthetic analysis of complex systems.

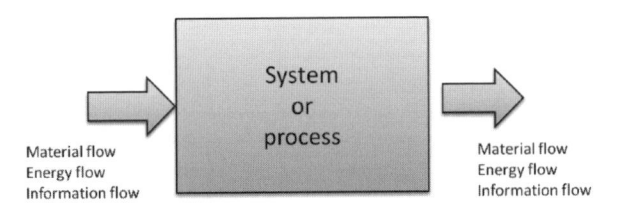

Figure 7.2. *Process model*

A system is represented graphically as a box, with arrows representing input and output. These arrows show the interactions between the system and other systems, or between the system and its environment. The environment may be seen as a system for which we only describe input and output. To list input and output elements, we consider:

– flows of material, in a broad sense, that is flows of material in liquid or gaseous form, flows of components, flows of people, etc.;

– energy flows of various forms;

– information flows that tell the system how to operate or which information it needs to use or process.

EXAMPLE 7.1.– Take the example of an automatic token-operated coffee machine. The first step consists of defining the system and its boundaries. A simple solution would be to choose the physical limits of the machine, the end of the electrical cable and the tap supplying the water. Cups are stored inside the machine. The phase of operations under consideration is the normal operational phase. A process model is shown in Figure 7.3. This modeling approach shows that it would also be useful to model the phase in which cups are supplied to the machine and tokens are retrieved, along with the machine maintenance phase.

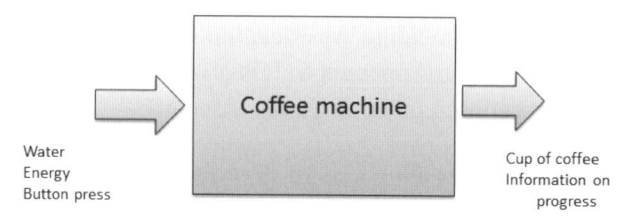

Figure 7.3. *Process model for a coffee machine*

7.2.2. *Hierarchical breakdown*

When modeling a system with a systemic approach, we often add a system representing the environment. This system acts as a supplier of input and as a consumer of output. This is shown in two parts (one on the right and another on the left) in Figure 7.4.

Moreover, when the system being analyzed is relatively complex, it may be represented as a number of subsystems which are a breakdown of the original system. The input and output elements of the initial system are distributed to the relevant subsystems. The interactions with the external environment are identical. An example of this type of breakdown is given in Figure 7.4.

7.3. Functional modeling

The objective of functional modeling is to describe what the system does, that is to allow us to identify the functions of a system (Figure 7.5).

DEFINITION 7.2.– *A function of a system is defined as "the action of a system expressed in terms of its purpose". The function is therefore a type of dematerialization of a system expressing its role, its actions, what it can do and how it behaves when faced with a need and the constraints imposed on it by the surrounding environment.*

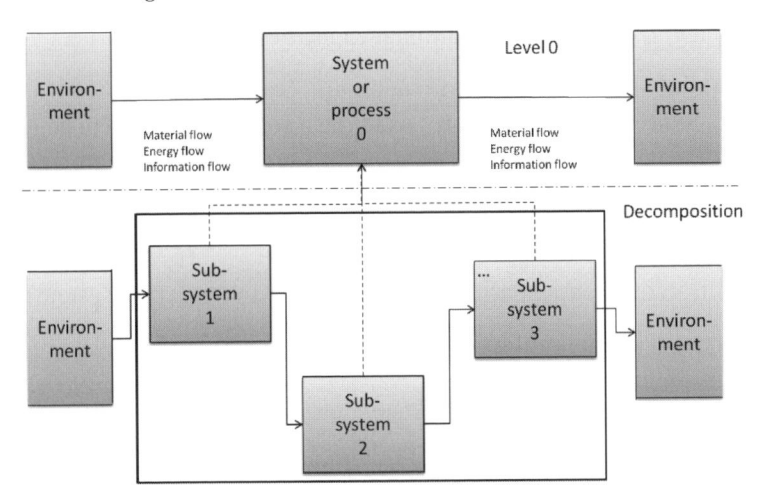

Figure 7.4. *Hierarchical breakdown*

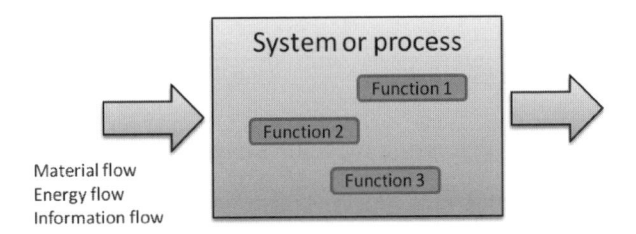

Figure 7.5. *Functions of a system*

7.3.1. *Identifying functions*

The standard approach to identifying functions consists of identifying the need or needs to which a system responds. It is then possible to identify several functions that respond to this need. The number of functions is limited to a reasonable value of the order of five, which may then, where necessary, be divided into subfunctions. We distinguish between:

– the main function(s) providing a response to the overall need, the main desire or the main requirement;

– technical functions, internal to the system, which allow the main function to be carried out;

– secondary functions that are not directly connected with the fulfillment of the requirement, but that give additional capacities;

– constraint functions: these functions are added to satisfy various constraints linked to operations – for example a sensor may need to be resistant to humidity.

The approach is divided as follows:

– definition of the boundaries of the object for analysis (the system):

- in terms of physical limits;
- in terms of operational phase;

– identification of main and secondary functions:

- to what need(s) does the system respond?
- what are the main stages of the process?

– identification of interactions with the exterior to identify constraint functions:

- what forms of aggression must the system be able to withstand?
- what operational and safety constraints does the system need to follow?

EXAMPLE 7.2.– The main function of a temperature sensor is to measure the temperature. The secondary function might be to trigger an alarm in the case of a loss of power; a constraint function might be a resistance to humidity.

Each function must be expressed in terms of a purpose, and should preferably be formulated using a verb in the infinitive with one or more complements specifying certain details of the function. As the aim of functional modeling is to seek potential dysfunctions capable of generating risks, it is important to precisely describe the functions and correctly specify conditions and operational constraints in order to avoid ambiguity.

EXAMPLE 7.3.– The specification of a heating function might be as follows: heat a recipient with a capacity of 10 L at a rate of 5°C/min.

7.3.2. *IDEF0 (SADT) representation*

In the course of operations, a function uses a certain number of entities or physical resources. These entities may take the form of materials, objects or energy that are transformed, or elements allowing the function to take place, such as machines, the workforce, methods or control information. The IDEF0 representation, presented in more detail in Appendix 6, allows us to define a function by showing these elements. It is then possible to place functions into a sequence, where the output of one function becomes the input, control data means or mechanisms of the next function.

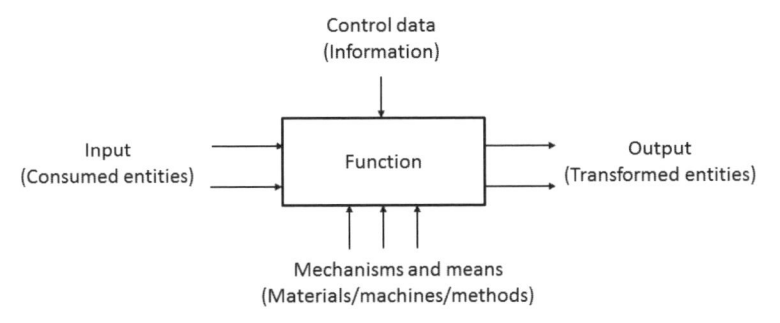

Figure 7.6. *IDEF0 function model*

A function may be divided into subfunctions. The performance of the main function involves the execution of some or all of the subfunctions of this main function. In this way, it is possible to create a hierarchical representation of the functions of a system.

7.4. Structural modeling

The aim of structural modeling is to describe how, and with what, a system is constructed. These entities are the elements that allow a system to carry out its functions. Structural modeling consists of creating a list of entities that make up the system.

Using the 5M method, the structural entities making up a system may be classified as:

– materials;

– machines;

– man power (actors);

– methods;

– measurement (information flow).

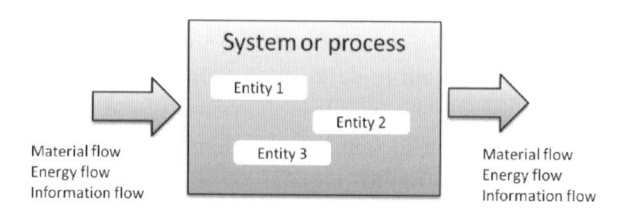

Figure 7.7. *Entities of a system*

These entities are used by the system or process to carry out its different functions or steps. To characterize the state of a resource, we may introduce a certain number of variables representing values and physical characteristics, such as temperature and pressure.

NOTE 7.1.– These elements are used in risk identification when seeking hazard sources according to their properties: chemical composition, chemical content, flammability, ignition sources, sources of electric current, radioactive elements, etc. Variables may be used to analyze deviations using the hazard and operability hazard and operability study (HAZOP) method.

Structural modeling is carried out using available descriptions of components and preexisting plans of the installation. From a practical viewpoint, a system may be seen as a vessel containing entities, and the input and output flows may be seen as entity flows. Certain entities may be used by several processes, and are said to be shared.

EXAMPLE 7.4.– Let us consider a production system made up of two batch production processes. The treatment time for each procedure is 6 h. The manufacturing process consists of loading materials, beginning treatment, transferring products and then cleaning the equipment. The two procedures operate with a time shift of 3 h and are run by the same operator. When modeling this production system, we could use two processes with a single human resource shared between the two.

In summary, the structural modeling approach to a system is as follows:

– Definition of the boundaries of the object being analyzed.

– Identification of elements:

- technical (materials and machines): technical elements correspond to the different physical elements encountered in an installation: machines, containers, contents, fluids, energy, etc.;

- human: most technical elements require physical actions carried out by an operator in the context of use (e.g. filling a reactor, feeding paper into a photocopier, etc.) or direction by an operator (switching on, information provision, etc.). They often require maintenance actions (such as changing the lamp in an overhead projector, cleaning a workspace, verifying a reactor, etc.). It is important to establish the most exhaustive list that may come out of these actions in order to identify the necessary human resources and validate the list of technical elements;

- organizational: the use of human and/or technical resources follows an operating mode or procedure within the framework of an overall organization. These aspects, which are not always explicit, make up the third category of elements requiring description;

- information: functions consume and produce information in the course of realization. It is important to identify the information that determines the operation of the system or of other systems.

– Choice of the system to which entity is assigned (with the possibility of sharing between systems).

– Analysis of interactions in the input and output of each system and identification of the entities linked to these flows.

EXAMPLE 7.5.– INDUSTRIAL SYSTEM: CHEMICAL REACTOR.– A system is shown in Figure 7.8. On the basis of this figure, analyzing the operation of the system, we are able to identify the following resources:

– Technical resources:

- tank (variable: V);

- exchanger and associated pipes (variable: T);

- stirrer blades;

- motor and shaft (variable: rotation speed);

- reaction mixture (variables: [C], T);

- cooling fluid (variable: T).

– Human resources:

- operators ensuring production in three 8-hr shifts (monitoring, loading and cleaning);

- the technician in charge of for adjusting captors and actuators;

- the worker in charge of for periodic verification of the reactor (corrosion checks).

– Organizational resources:

- the production recipes;

- emergency procedure in case of reactor overheating;

- maintenance procedures;

- production plans.

– Information resources:

- production order;

- production report.

7.5. Structuro-functional modeling

Structuro-functional modeling is a combination of the two approaches described above. The installation being analyzed is divided into systems or processes; note that a process or system is seen as an organized set of activities that uses physical entities (workforce, equipment, material and machines, raw materials and information) as resources to transform input elements into output elements, with the final goal of creating a product. Each process (Figure 7.9) is described internally in terms of:

– its functions;

– the entities involved and the entities it processes;
– the interactions between these functions and entities;
– interactions with other processes.

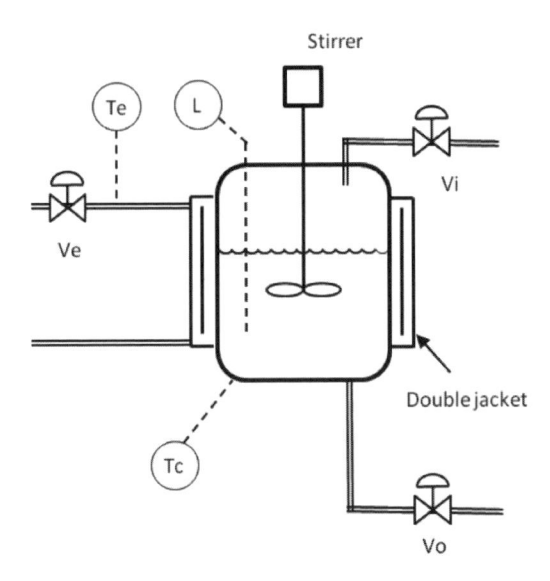

Figure 7.8. *Example of a chemical production system*

The entities may be input, output or control elements or function mechanisms (Figure 7.10). They are attached to the system that uses them, the system that produces them or the system that consumes them. When the entity and the function do not belong to the same system, an entity flow appears. An entity may be shared by several systems; this is the case when an operator carries out tasks for different processes.

The model may be presented in graphical form, or in a tabular form, as shown in the examples below.

EXAMPLE 7.6.– ORGANIZATIONAL SYSTEM: ALERT MANAGEMENT IN A CITY EMERGENCY PLAN.– A city emergency action plan includes an alert management phase. The purpose of this phase is to alert the population using a mass alert system when a phenomenon concerning civil security occurs. The diffusion of the alert may be modeled by a process for which the main function is to deal with the alert better; the resources used are shown in Figure 7.11. The input and output flows are made up of entities.

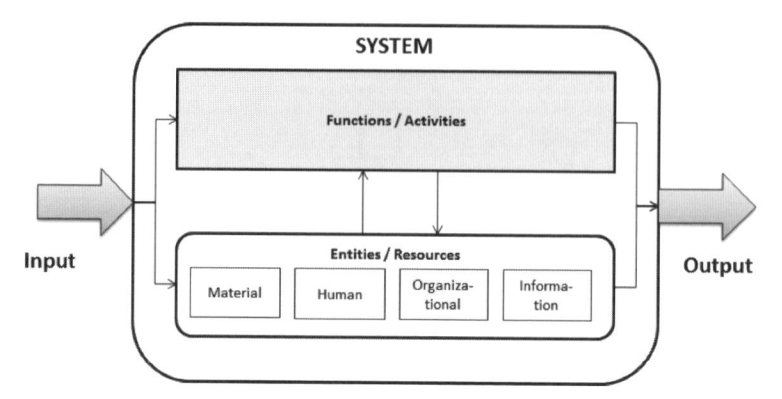

Figure 7.9. *Model of a system*

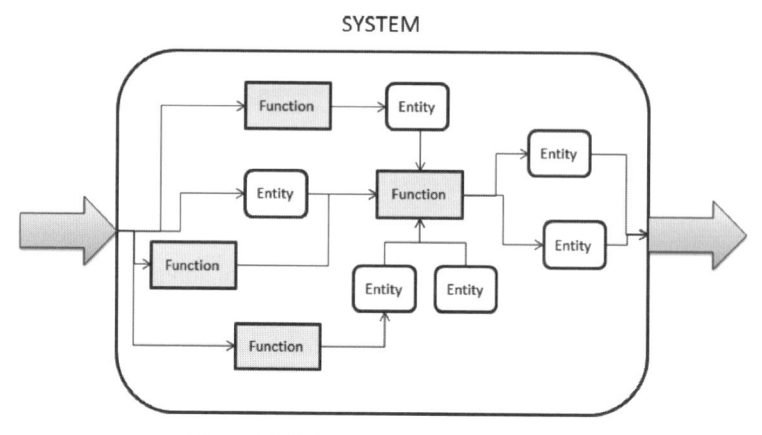

Figure 7.10. *Decomposition of a system*

EXAMPLE 7.7.– DESK LAMP.– Let us consider the desk lamp shown in Figure 7.12. The lamp has a 220 V power supply and is intended to light a desk. It is often plugged in and unplugged without particular precautions being taken, and the cable shows signs of wear near the plug. The lamp uses an incandescent bulb. The electrical installation is recent and includes a differential circuit breaker. The user is responsible for minor maintenance tasks, such as changing the bulb. The structuro-functional model of the system is shown in Figure 7.13.

Modeling of Systems for Risk Analysis 137

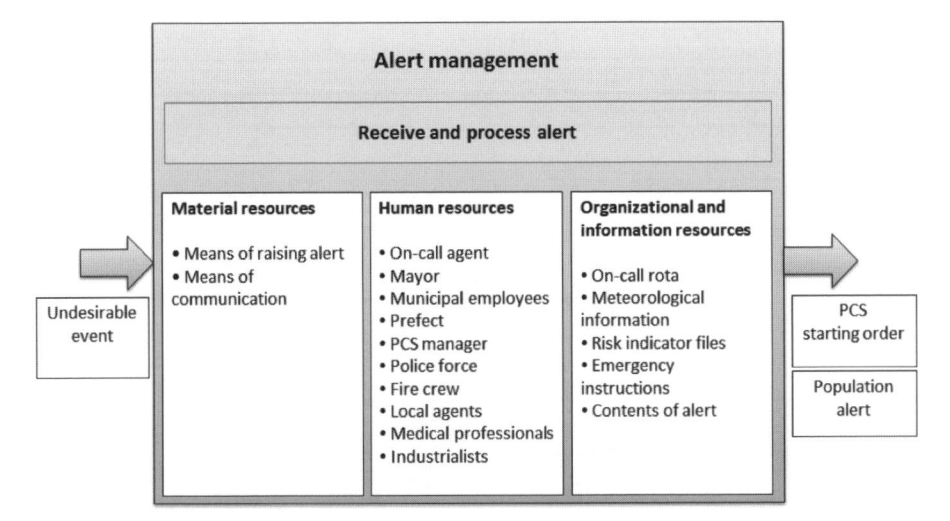

Figure 7.11. *Example of structuro-functional modeling*

Figure 7.12. *Desk lamp*

7.6. Modeling the behavior of a system

In addition to a description of what a system "does", we may wish to describe the sequence of different activities that it carries out. This model of the evolution of a system is known as a system behavior model. It allows us to introduce notions of the order in which actions should be carried out, conditional choices of actions based on the state of the system and the result of previous actions, and a description of temporal aspects (duration of actions, delays, etc.). This model is much more detailed than those described above, and its construction requires considerably more effort; for this reason, its use in risk analysis is limited to certain specific cases.

138   Risk Analysis

System	Function	Structural entity
Lamp	Produce light	
Lamp	Direct light flow	
Lamp		Light bulb
Lamp		Arm
Lamp		Base
Lamp		Light shade
Lamp		Switch
Lamp		Plug
Lamp		Cable
User	Switch on/off	
User	Direct	
User	Plug in/unplug lamp	
User	Change light bulb	
Environment	Protect against electrical hazard	
Environment		Circuit breaker
Environment		Power supply
Environment		Office
Environment		Light produced

Figure 7.13. *Tabular model of the lamp*

Control
- Start/stop order

Input
- Electrical power

Produce light

Output
- Light produced

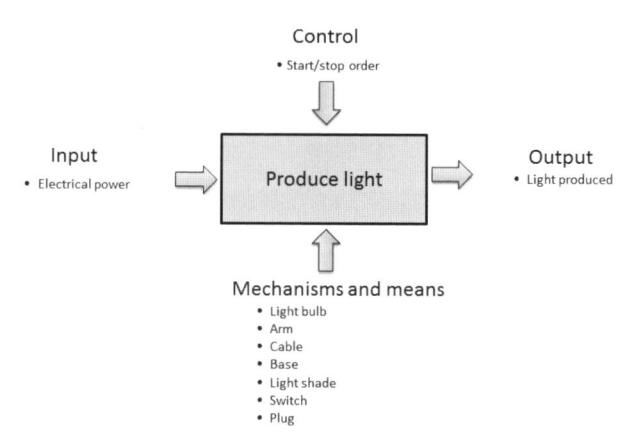

Mechanisms and means
- Light bulb
- Arm
- Cable
- Base
- Light shade
- Switch
- Plug

Figure 7.14. *IDEF0 model of the function "produce light"*

The activity diagram used in the Unified Modeling Language (UML) method, discussed in Appendix 6, provides a suitable basis for describing the behavior of a system in formal terms. This diagram is constructed using a small selection of geometric shapes (Figure 7.15):

– Rectangles with rounded corners representing activities.

– Diamonds representing decisions.

– Bars representing the start (*split*) or end (*join*) of parallel activities.

– A full black circle represents the starting state, and a circle with a black center represents a final state.

– Directional edges between these elements describe the flow of activities and the order in which they take place.

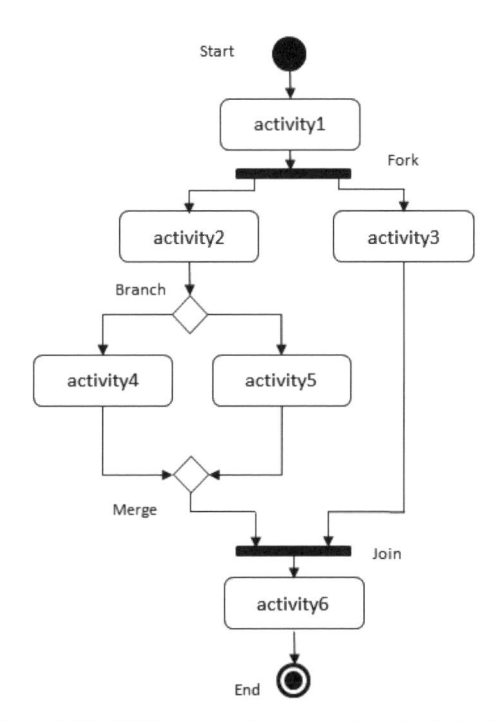

Figure 7.15. *UML activity diagram to describe behavior*

This modeling approach represents sequences of activities in a similar way to classic flow charts: diamonds are used to represent alternatives and to bring elements together. It allows us to add a behavioral description aspect to the structuro-functional approach presented above. Activities may be seen as functions or subfunctions identified in previous stages. The different variables used in decision blocks are those of the functions and entities of the system.

Behavioral modeling allows us to analyze errors that may be due to bad sequences of activities, which may still have been carried out correctly at an individual level, or errors linked to a wrong decision leading to the execution of the wrong activity.

EXAMPLE 7.8.– An example of this type of model is shown in Figure 7.16. This example describes the management of an alert in an emergency plan.

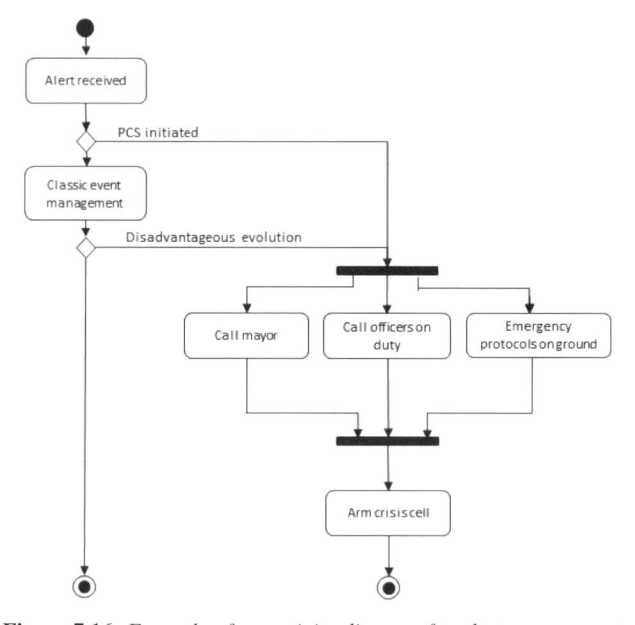

Figure 7.16. *Example of an activity diagram for alert management*

7.7. Modeling human tasks

Most risk analysis methods for human tasks use a modeling of the tasks to be carried out. This modeling then acts as a basis for identifying potential human errors. Specific modeling methods have been developed for this context, although they are based on the same concepts discussed earlier in this chapter.

We will now present three approaches used for implementing methods for the analysis of human reliability, which will be discussed further in Chapter 14. These methods concern the modeling of operator activity. This term designates the set of actions carried out by a user to accomplish a task.

7.7.1. *Hierarchical task analysis (HTA)*

This approach provides a structured means of representing tasks that must be accomplished. It is used in a certain number of applications where it is necessary to provide a systematic description of the task of a human operator, such as interface design, the creation of user manuals or procedures, training and, evidently, risk analysis processes concerned with the potential for human error.

HTA is a top-down approach first proposed by Annett *et al.* [ANN 67, ANN 00]. The key principle consists of dividing an overall task into subtasks, then, if necessary, into elementary tasks that may themselves be divided further as required. In parallel, an action plan is established to describe the sequence of these subtasks. The analyst must examine the way in which the operator's actions occur, and, where necessary, talk to the operator in order to identify action plans for attainment of the desired goal.

Type of plan	Example
Linear	Implement 1, then 2, then 3
Non linear	Implement 1, 2 and 3 in any order
Parallel	Implement 1 and 2 simultaneously
Conditional	Implement 1. If result not obtained, implement 2 otherwise 3
Cyclical	Implement 1, 2 and 3 then repeat

Figure 7.17. *Examples of plans associated with tasks*

The HTA method includes the following steps:

– Definition of the task to analyze: this is the first step. The aim of this task, the context in which it is carried out, presupposed conditions and the limits of the mission must be specified. If we consider the task "make coffee", for example, we must define the aim, the raw materials and the equipment available for making coffee.

– Collection of useful information: we need to collect information on the stages of the task, the technology used, man–man and man–machine interactions, decisions that need to be taken and the constraints of the tasks. Several possible approaches exist for data collection: observation *in situ*, interviews, questionnaires, visits, etc. The choice of a method depends on the time available and the accessibility of the site.

– Definition of subtasks: using the gathered information, the task is divided into subtasks (generally from four to six, but this is not obligatory).

– Decomposition of subtasks: the subtasks identified in the previous stage are themselves divided into subtasks. The process is repeated until the desired level of decomposition is reached. Two levels of decomposition are usually sufficient. The basic task used in an HTA analysis should be a relatively simple operation that may be carried out directly by an operator.

– Construct a plan of the task: during this stage, we define the sequence of different subtasks, potential alternatives and conditional sequences.

EXAMPLE 7.9.– Take the example of making coffee using an electric coffee machine. The overall task is to prepare coffee. The ingredients are presumed to be available and close by, as are the cups. The main task is divided as follows:

1) load coffee;

2) load water;

3) switch on and wait;

4) pour coffee.

The plan associated with this decomposition is linear: perform 1, then 2, then 3, and then 4.

These subtasks may be divided further. For subtask 1, we obtain:

i) open the filter receptacle;

ii) check for used filter;

iii) dispose of used filter;

iv) install new filter;

v) add coffee powder;

vi) close filter receptacle.

The plan associated with this decomposition is as follows:

– perform i and ii;

– if filter is present, then perform iii,

– perform iv, v and vi.

Modeling of Systems for Risk Analysis   143

The level of detail used in the decomposition should be chosen based on possible potential errors: the performance factors[1] associated with the elementary task (training, difficulty level, conditions of realization, etc.) must be relatively homogeneous and the possible errors should only lead to one type of consequence.

There are two main ways of representing an HTA, either in diagrammatic or in tabular form. The two approaches are complementary: the first representation provides a synthetic overview, while the second representation provides a higher level of detail.

An example is shown in Figure 7.18. The figure may also be represented in a tabular form, using the format presented in Figure 7.19. Column 1 contains an enumeration of tasks, column 2 describes the aim or actions of the task, column 3 describes the conditions to begin execution of the task, column 4 describes the expected result from the system on which we are acting and column 5 shows the duration or other temporal aspects of the task.

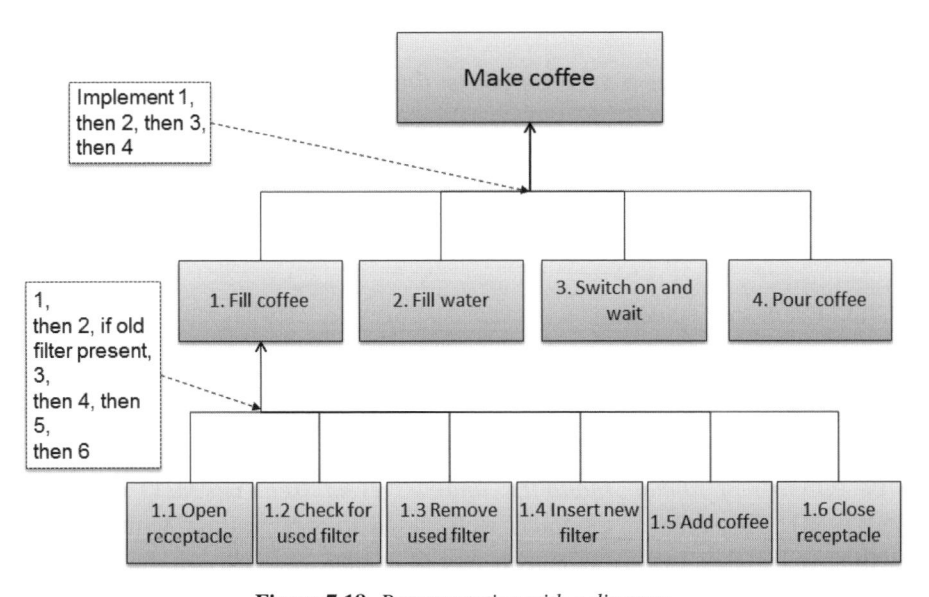

Figure 7.18. *Representation with a diagram*

---

1 See Chapter 14 for details of this notion.

HTA modeling is an interesting tool that presents a certain number of advantages. The method allows us to create an organized representation of the tasks of an operator, up to the required level of detail. It also allows collaborative working with different personnel. It does, however, have certain limitations: the method is time consuming in the case of complex tasks, and is not suited to describing the cognitive aspects of the task being analyzed. This aspect will be discussed further in Chapter 14.

N°	Description	Conditions	Effects of actions	Temporal aspects	Comments
(1)	(2)	(3)	(4)	(5)	(6)

Figure 7.19. *Tabular representation*

7.7.2. *Modeling using a decision/action flow diagram*

This approach is similar to the UML activity model discussed in section 7.6 and used to model the behavior of a system. An example is shown in Figure 7.20.

7.7.3. *Event tree modeling*

In this approach, an event tree is used to represent the sequence of different actions and decisions that an operator should carry out when faced with an undesirable event. These diagrams are known as *operator action event tree* (OAET) or *human reliability analysis event tree* (HRAET) diagrams [SWA 83]. This type of model is well suited to representing sequences of actions to be taken by an operator in case of an undesirable event, for example in an emergency situation. A detailed presentation of event trees is given in Chapter 13. Figure 7.21 shows an example of an OAET diagram, modeling the actions expected of an operator in the case of a temperature deviation.

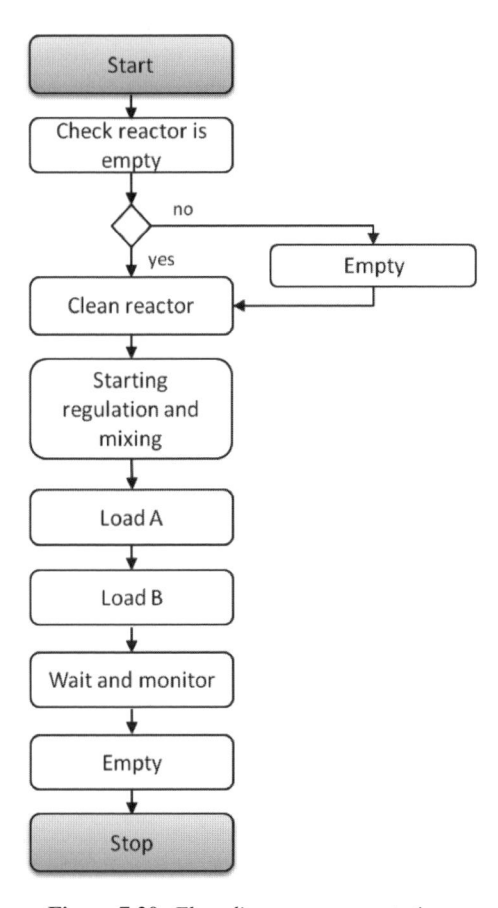

Figure 7.20. *Flow diagram representation*

7.8. Choosing an approach

The methods presented in this chapter are not mutually exclusive or opposed, but instead constitute complementary approaches. When selecting a method, the key consideration is to choose a suitable level of detail. Figure 7.22 shows a progression from the systemic to the structuro-functional approach. A behavioral model could also be added.

As a first step, systemic modeling allows us to describe the object of study or the system to analyze. This type of modeling is sufficient for risk identification and preliminary risk analysis (PRA). It is also suitable for use with a systemic approach, such as Systemic and Organised Risk Analysis Method (SORAM). If certain parts of the system need to be analyzed in greater detail, we then construct a functional, structural or structuro-functional model to implement an Failure Mode and Effect Analysis (FMEA)- or HAZOP-type approach. A behavioral description may be envisaged in cases requiring a study of human reliability or analysis of human barriers.

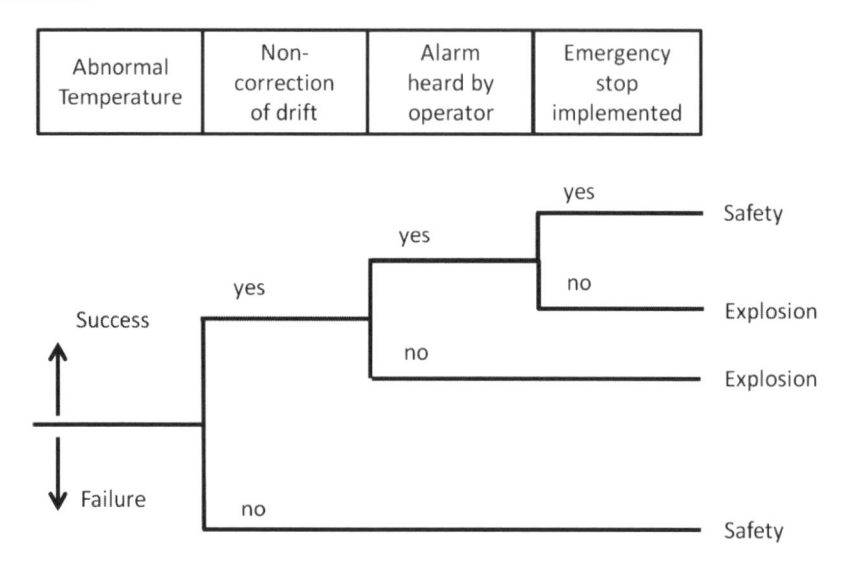

Figure 7.21. *Event tree of operator actions*

The whole approach is carried out in an iterative manner: the model may be added to, modified or corrected in the course of the risk analysis process.

7.9. Relationship between the system model and the risk model

System models are used to structure risk analysis. The elements used to describe scenarios allow us to describe undesirable events or faults related to different elements of the system model. If a systemic modeling is used, all events are associated with systems (Figure 7.23). If the model is more detailed, showing functions and structural entities, events may either be associated directly with system levels or with functions and entities (Figure 7.24).

Modeling of Systems for Risk Analysis 147

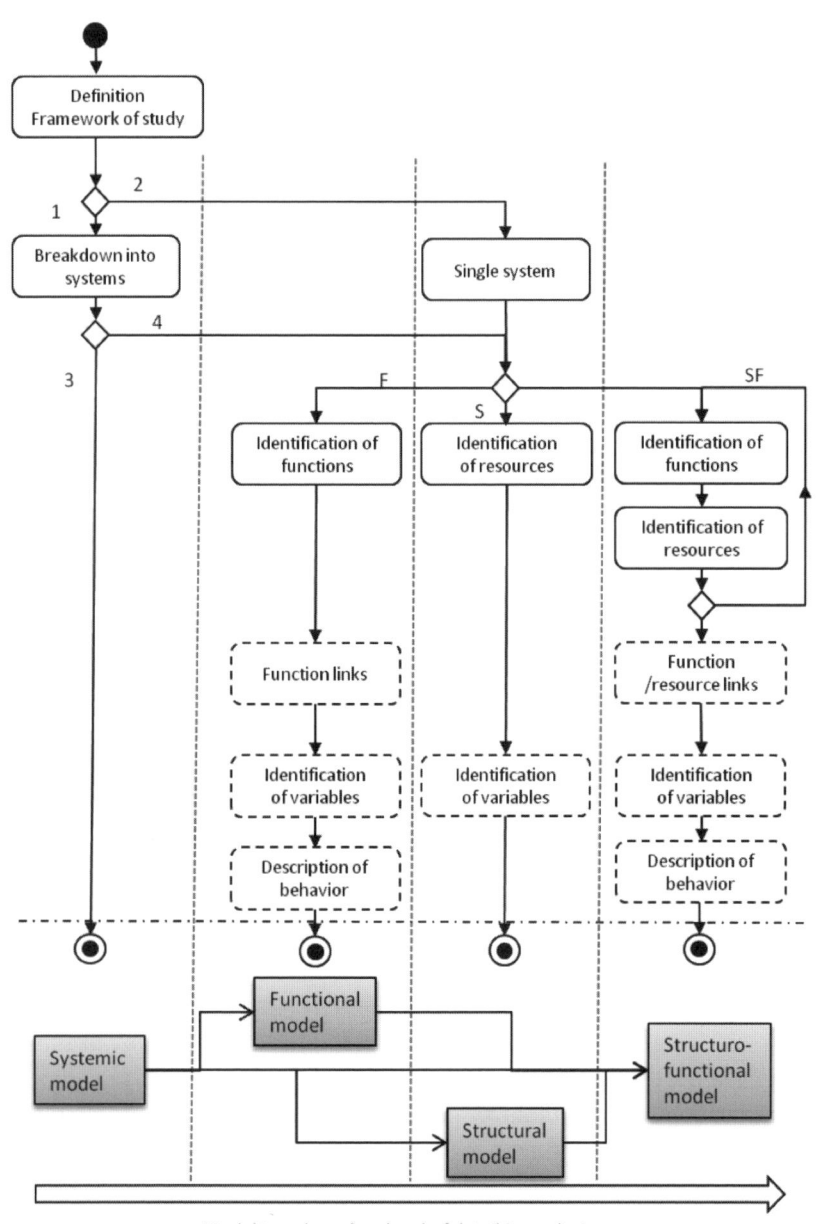

Figure 7.22. *Modeling approach according to needs*

148   Risk Analysis

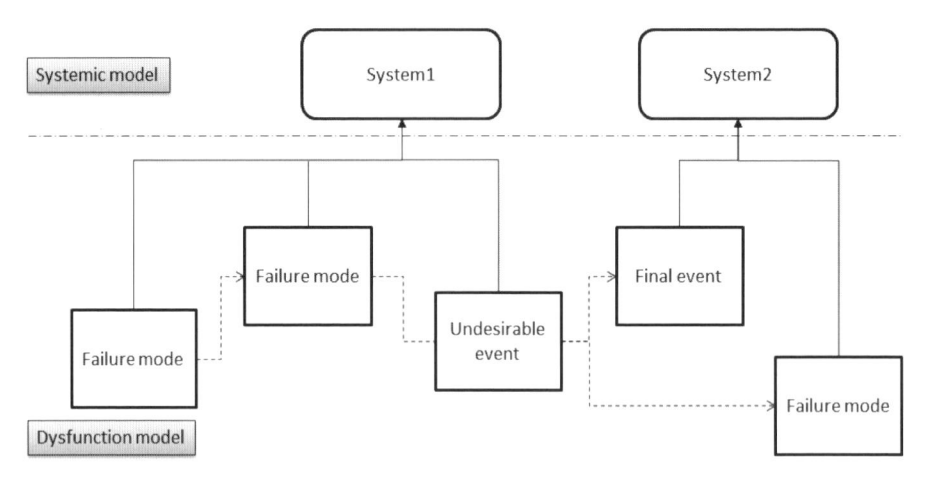

Figure 7.23. *Links between the systemic model and risk analysis*

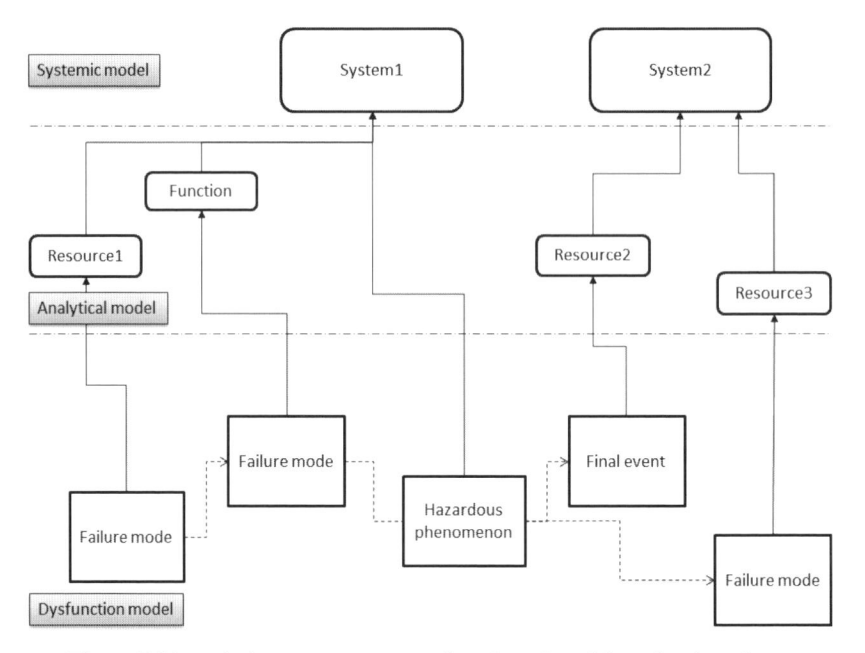

Figure 7.24. *Links between systemic and analytical models and risk analysis*

These events may be classified into different, non-exclusive, categories:

– Central undesirable event: this type of event is used to identify the events at the center of scenarios. It may be associated with a system, a function or a structural entity. These events correspond to the lines in the table of the PHA method and to the center of bow-tie scenarios.

– Hazardous phenomenon: this type of event characterizes a hazardous phenomenon, generally associated with a system. Events of this type are used to describe hazard flows using the SORAM method.

– Failure: this event type characterizes a fault or failure, and instances of this sort are identified in the framework of the FMEA method.

– Variable deviation: this type of event describes deviations in variables, as identified by the HAZOP method.

– Degradation: this type of event characterizes damage. They may be associated with a system at global level, or, in a more precise manner, with a function or a structural entity. These events are those characterized by a severity value. They generally correspond to the final events of scenarios.

– Intermediary abnormal event: this group of event corresponds to events that are useful in describing a scenario, but that do not fall into any of the other categories described above.

Using this classification, the different risk analysis methods may be seen as approaches to the construction of a global model of system dysfunction. The use of a single model facilitates the analysis and the move from one method to another, as we will see in the chapters describing these different methods.

PART 3

Risk Analysis Methods

Chapter 8

Preliminary Hazard Analysis

8.1. Introduction

Preliminary hazard analysis (PHA) was first used by the US army in the 1960s and was described in the first version of the MIL-STD-882 standard [DOD 12] in 1969. It was then adopted by a certain number of industries, notably in the aeronautical, chemical and nuclear fields. In France, the Union of Chemical Industries (UIC) published a methodological guide designed for the chemical industry and its use was recommended from the early 1980s.

The purpose of this method is the *a priori* identification of hazardous undesirable events, that is identification before an accident can occur. To do this, we review the elements present in an installation and the activities involved to look for the sources of danger and examine the possibility of occurrence of undesirable events. For a system, we examine the components and operations, and, where necessary, the possible damaging consequences. The importance of the different identified risks is generally assessed in a qualitative or semi-qualitative manner.

This method may be applied in a global manner without requiring detailed knowledge of the installation concerned. It constitutes a first step taken before starting a more detailed analytical process (Figure 8.1). The term "preliminary" means that the method may be used:

– as part of a preliminary phase in the development of a new system, at design stage, when little information is available: the aim of the process is to identify potentially hazardous situations as early as possible in order to either modify the design or implement appropriate risk control techniques from the outset;

– as a preliminary stage in risk analysis, to identify risks and determine their importance. For a simple installation, or if the risk level may be considered acceptable, the risk analysis process may stop at this point. In the case of a complex installation, or if the observed risk level is not acceptable, a more detailed method of analysis, such as Failure Mode and Effect analysis (FMEA), Hazard and operability analysis (HAZOP) or a fault tree, will be used.

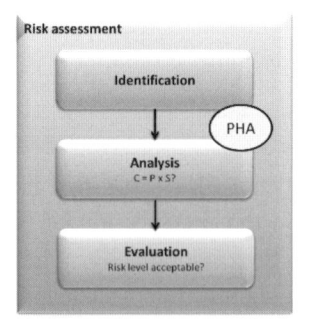

Figure 8.1. *PHA in the risk assessment process*

This method is described in the MIL-STD-882 standard, but a number of variations exist, producing results tables of different forms. In cases where the likelihood and the severity of the risk are not assessed, we speak of a preliminary hazard list (PHL).

The MIL-STD-882 standard has undergone a series of revisions since the publication of the first version in 1969, to take account of evolutions in risk analysis and management. Version A was published in 1977, B in 1984 (with corrections in 1987), C in 1993 (corrected in 1996), D in 2000 and E in 2012. The standard has, therefore, kept up with changes and its recommendations remain suited to current needs.

The PHA method, described briefly in the initial versions of the standard, was not included in version D, published without the method sheets included in previous versions. A more detailed description of the method was reintegrated into version E. The definition of scales and risk matrices has also evolved over the course of different versions:

– MIL-STD-882: definition of severity levels, no risk matrix;

– MIL-STD-882A: definition of severity levels and qualitative probability, no risk matrix;

– MIL-STD-882B: qualitative risk matrix;

– MIL-STD-882C: qualitative and quantitative risk matrices, acceptability levels;

– MIL-STD-882D: qualitative risk matrix, quantitative probability levels, non-detailed methods;

– MIL-STD-882E: qualitative risk matrix using levels with quantitative equivalents, introduction of "eliminated" probability level.

In this chapter, we will present the version described in revision E, which is the current version of the MIL-STD-882 standard, along with different variants of PHA.

8.2. Implementation of the method

The general PHA approach is shown in Figure 8.2. The real system is first observed, and then described. This description must contain useful information on the elements that make up the system and on the activities of the system. Using this description, the analyst or study team then identifies hazards based on checklists, expertise or feedback, and/or through brainstorming.

At this stage, we obtain a list of hazards. Each hazard is then associated with the undesirable event or events that relate to the occurrence of the hazard, often specifying the associated hazardous situation. We then analyze possible damage, before evaluating the likelihood, and then the severity of the risk. The PHA approach is thus made up of the following stages:

1) Analysis preparation: definition of context, gathering information and observations.

2) Description and modeling of the installation.

3) Identification of hazards and undesirable events.

4) Analysis of hazardous situations.

5) Analysis of consequences.

6) Search for existing barriers.

7) Evaluation of severity and frequency or likelihood, using a qualitative or semi-qualitative approach.

8) Proposal of new barriers (optional).

9) Creation of the report about the analysis.

We will now consider these stages in detail.

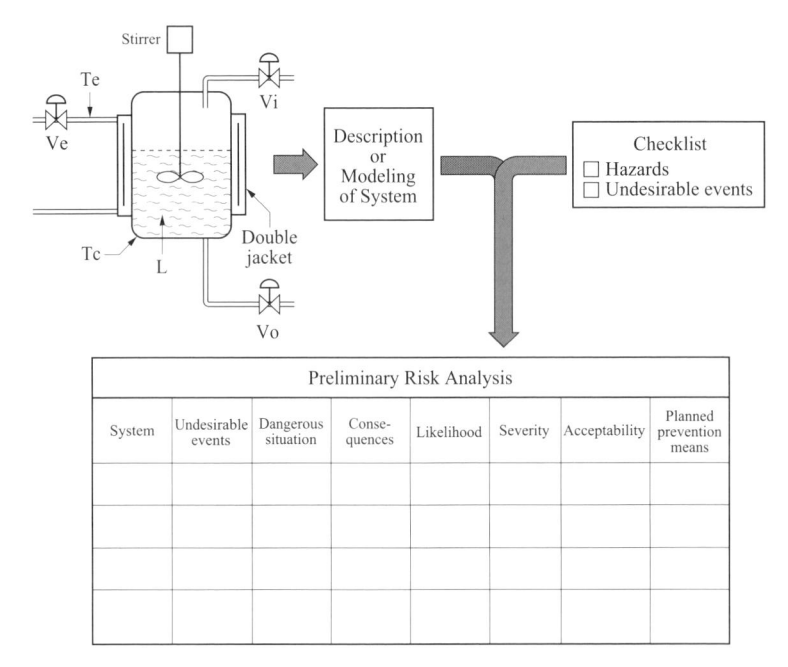

Figure 8.2. *Overview of PHA approach*

8.2.1. *Definition of context, information gathering and representation of the installation*

Before beginning a PHA, we need to define a precise framework. We must specify:

– the type of PHA: the degree of analysis, the format of the results table and the list of risks to examine. These aspects are described below;

– the approach chosen to measure the risk (Chapter 6) and the way in which an acceptable level is defined. A qualitative approach is generally chosen, and acceptability levels are defined using a risk matrix.

The following stage consists of gathering useful information. As PHA may be carried out relatively early in the design cycle, the data available for the installation in question may be somewhat limited. In this case, we use plans of the installation or system, details of different elements, the planned mode of operation, experimental feedback and information from other similar installations. In cases where the installation exists, we may visit the site or examine the object in question.

This information is then formatted to give a complete description of the installation. In a general manner, we identify different structural elements, and we usually consider the main function and activities of the installation. This task may be carried out by dividing the installation or system into zones or subsystems that are easier to process. For each system or sub-system, we wish to identify:

– the entities (material, machines or equipment, man power and methods) involved;

– the functions or activities of the system.

In this way, we build a structural or structuro-functional model of the installation, as described in section 7.8, which may be represented by a table as shown in Figure 8.3.

In the case of systems with significant dynamic aspects, a behavioral model may be produced, for example, using the approach presented in section 7.6. This model is used to structure the analysis of risks linked to certain functions that are accomplished correctly, but with insufficient temporal performances, and risks caused by activities carried out in the wrong order or incorrectly chosen.

NOTE 8.1.– A description of the environment using the 5M approach may be produced by adding an additional system, known as the "system environment", containing external elements in interaction with the object of study, which need to be taken into account in risk analysis.

N°	UNITS or SUB-SYSTEMS	ENTITIES What the system is made from	FUNCTIONS/ACTIVITIES What the system does

Figure 8.3. *System model in the form of a list of elements and activities*

NOTE 8.2.– In the remaining chapter, we will use the term "system" to designate the object, installation or organization being analyzed. The term "element" will be used to denote the entities, activities or functions that make up the system.

8.2.2. *Identification of hazards and undesirable events*

Using the representation of the system, the identification stage consists of creating an inventory of hazardous entities and activities that may produce a hazardous undesirable event. This event may be generated by a single element, or by an undesirable interaction between different elements. These elements may include:

– hazardous equipment or machines, such as bulk stores, machines with moving parts, elements under pressure, etc.;

– hazardous substances, mixtures or fluids used in different stages of a manufacturing process;

– unsuitable, undefined or poorly followed methods;

– badly trained employees;

– potentially hazardous activities, such as mechanical handling or hazardous manufacturing operations;

– functions carried out in a way not in accordance with specifications;

– badly designed equipment or processes, etc.

Domain-specific guide lists are generally used in identifying these elements and activities. The general form of this type of list is shown in Figure 8.4. A generic list of different hazards is shown in Table 8.1. Obviously, lists of this type must be adapted depending on the type of system and domain of activity.

LIST OF HAZARDS

Hazard	Undesirable event	Type of entity or activity concerned (optional)
Electrical hazard	Electric shock, electrocution	Live element
....	....	....

Figure 8.4. *Format of the hazard table*

Each element of the system is analyzed using this list of potential hazards. The result of the analysis may take the form of a system elements/hazards matrix (Figure 8.5). For each hazard, we identify the associated undesirable event, allowing us to specify the potential risk, unless we only require a very synthetic level of hazard analysis.

			\multicolumn{9}{c}{HAZARDS}
			Fire	Thermic	Mechanical Machines	Electrical	Noise and vibrations	Chemical	Biological	Non-ionizing radiation	Ionizing radiation
ENTITIES & ACTIVITIES	S1	Entity #1					x			x	
		Entity #2		x							
		....									
		Activity #1						x			
		...									
	S2	Entity #1	x		x						
		...									

Figure 8.5. *Hazard identification table*

At this stage in the analytical process, it is useful to remember Murphy's famous law: what can go wrong, will go wrong. Any risk that is not taken into account at this stage will not be treated later in the process. However, if a risk identified during this stage proves to be unimportant, this will be seen at a later stage when analyzing it and assessing its likelihood and severity.

EXAMPLE 8.1.– If the system under study includes electrical equipment, for example, there is a potential electrical hazard, and the associated undesirable event takes the form of an electric shock.

8.2.3. *Analysis of hazardous situations, consequences and existing barriers*

Once the hazard and the undesirable event have been identified, the following step consists of examining the scenario S that describes the risk. For each undesirable event, we specify:

160   Risk Analysis

– the damage affecting the potential targets;

– the conditions in which the undesirable event may occur.

The scenario of occurrence of a risk is shown in Figure 8.6, highlighting the different types of event used to describe the sequence:

$$Dangerous\ situation \rightarrow Undesirable\ event \rightarrow Consequences$$

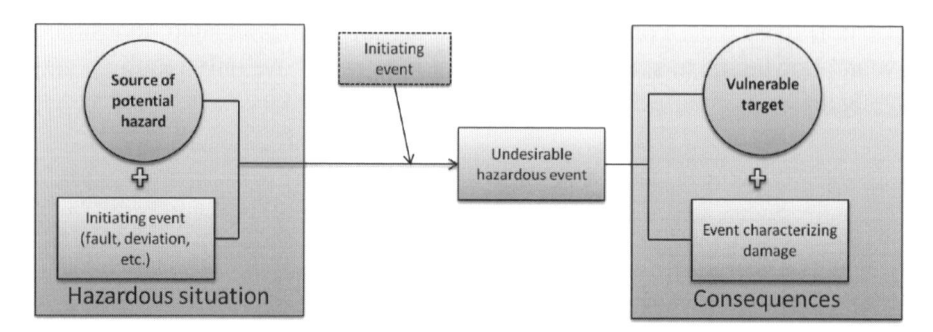

Figure 8.6. *Scenario leading to damage*

The undesirable event characterizes a mechanism or phenomenon that generates damaging consequences. It takes place in a particular situation, known as a hazardous situation. When an element of a system is in this situation, the undesirable event may occur either spontaneously, given the information and system model we have, or after the occurrence of a particular event that can happen in the course of normal operations. The event, which characterizes the beginning of occurrence of the undesirable event, is known as the "initiating event". Identification of this event may be useful in specifying the conditions of occurrence of the undesirable event, but it is optional.

EXAMPLE 8.2.– Examples of scenarios are shown in Figure 8.7. In the case of scenario 1, the trigger event might consist of switching on the machine. For scenario 11, the hard drive failure is considered to occur spontaneously. This failure may, in fact, be the result of a slight deviation from the temperature range compatible with the operating specifications of the hard drive, but this event is not modeled.

Preliminary Hazard Analysis 161

A hazardous situation may be described using the source of potential hazard and a deviation:

– The source may be an entity, a piece of equipment, a method, a person or an activity.

– The deviation is an item of information concerning the state of the source, describing the specific conditions in which an undesirable event may occur. This deviation may be a fault appearing in the source, or in the conditions of use in cases of an entity, or in implementation in the case of an activity. This information concerning the deviation is used to assess the likelihood of the hazardous situation appearing. In certain cases, deviation information is not specified, and it is implied that the entity or activity might generate an undesirable event in the course of normal usage or implementation conditions. In other cases, only the deviation is specified as the element in question is implied.

N°	Hazardous situation	Undesirable event	Consequences
1	Electrical aparatus with poorly insulated areas	Electrocution	Burns, cardiac arrest, death
2	Rotating equipment with insufficient protection	Operator caught in mechanism	Injury
3	Fuel vapor close to an ignition source	Explosion	Ear drum problems, burns, death, loss of equipment
4	Use of toxic products	Intoxication	Respiratory disease
5	Nitrogen leak	Lack of oxygen	Anoxia
6	Poorly propped earthworks in a trench	Collapse and burial	Injury, death, material losses
7	Steam heater with blocked security valve	Explosion due to overpressure	Injury, death, material losses
8	Poor handwashing between patients	Contamination	Nosocomial disease transmission
9	Pill boxes filled in stressful conditions	Medication error	Iatrogenic disease
10	Lack of way markers in project	Undetected drifts in relation to objectives	Final objectives not met
11	Work not saved periodically	Hard disk failure	Loss of data
12	Poor mastery of emergency plan by actors	Poor organization of emergency services	Loss of life

Figure 8.7. *List of hazards and associated undesirable events*

EXAMPLE 8.3.– In scenario 1, the entity in question is a piece of equipment carrying an electrical current, and the fault is "poorly insulated". For scenario 4, the deviation is not specified as the activity itself is considered to be hazardous. In case 11, the activity is not specified, implying that it refers to the use of a computing system.

The undesirable event generates damaging consequences for various targets. Depending on the type of PHA, we may look at human, material, environmental or other types of target. The consequences of each undesirable event may be described by specifying the target and providing information that characterizes damage. In a number of cases, the target is implied, and only the type of damage is described.

EXAMPLE 8.4.– In scenario 1, the target is the operator; this is not specified explicitly.

Once a hazardous situation has been analyzed, we must identify existing prevention and protection measures. Note that, in a certain number of cases, the hazardous situations identified are actually linked to the insufficiency of these barriers.

In summary, the process of analyzing hazardous situations and their consequences takes place in the following manner:

– Selection of the central undesirable event.

– Identification of the associated hazardous situation(s):

　- the element that is the source of the hazard;

　- deviations in the state of this element.

– Optional examination of the cause of deviations.

– Analysis of the consequences associated with the central undesirable event:

　- elements subject to consequences;

　- degradation or changes in the state concerning these elements.

– Analysis of the existing barriers.

This information is gathered in a table, with a row for each central undesirable event (Figure 8.8). In certain cases, the consequence of an undesirable event is another undesirable event: a fire may lead to an explosion, for example, producing a domino effect. This is described by several rows in the PHA table.

Preliminary hazard analysis												
System	Causes of hazardous situation	Hazardous situation: element and drift	Undesirable event	Consequences: targets and damage	L	S	A	Existing barriers	L'	S'	A'	Barriers to add	L"	S"	A"
	optional				optional		optional

Figure 8.8. *Standard PHA table*

8.2.4. *Assessment of severity and frequency or likelihood*

The likelihood and severity of risks are assessed using a qualitative or semi-qualitative approach. The scales and risk matrix used are defined in the context definition phase. Sections 6.2 and 6.3 present the approaches used in defining these scales.

The likelihood of an undesirable event is assessed by examining hazardous situations and their possible causes. The severity of the event is determined by examining its consequences. In cases where the consequences do not correspond to real damage, but rather to a change in state that provokes an undesirable event, which then creates damage, we need to examine the final consequences in the chain in order to evaluate the severity. This is the case, for example, of electrical faults, which may be the cause of a loss of control; this loss of control then leads to damage.

Likelihood and severity are generally evaluated with and without existing barriers in order to highlight the initial, without safety barriers, risk level. For this reason, we add columns to the table (Figure 8.8) describing barriers and the associated risk levels.

The risk level is determined using a risk matrix (section 6.4).

8.2.5. *Proposing new barriers*

Once the level of risk associated with each undesirable event has been evaluated, the final stage consists of proposing safety barriers to make this risk acceptable. These barriers may act on different levels:

1) by limiting the occurrence of hazardous situations;

2) by reducing the possibility of occurrence of undesirable events from the hazardous situation;

3) by limiting the consequences of the hazardous event;

4) if the causes of the hazardous situation are known, we can operate earlier in the process, limiting the occurrence of these causes.

Cases 1, 2 and 4 are preventive measures, whereas case 3 is a protective measure. Barriers are added to the table using an extra column (Figure 8.8), and the risk level is re-evaluated taking account of these barriers.

EXAMPLE 8.5.– In scenario 1, the reinforcement of preventive maintenance would limit the occurrence of badly isolated elements. To limit the possibility of contact, we might wish to prevent direct access to the machine in question. Users might be required to wear personal protective equipment to limit the effects of passage of an electric current, or an earth plug, and a differential circuit breaker might be added.

8.2.6. *Limitations*

PHA constitutes an efficient method for risk analysis, but it also has certain limitations:

– PHA is based on checklists, which must be suited to the type of system.

– While the method appears simple, it requires considerable rigor on the part of the analyst, as the clarity and effectiveness of analysis depend on the coherency of the approach. The following specific considerations are important:

- the level of detail of the examined elements must be homogeneous;

- the identified scenarios must be consistent with each other; an event must be designated in the same manner in all scenarios, and sequences of events must be made clear.

– For complex scenarios, a single row in a table is not sufficient for complete description, both in terms of causes (combination of events and sequences) and effects (different possible effects and possible event sequences): a bow-tie-type diagram is preferable in these cases.

For this reason, PHA studies are often followed by a more detailed study for critical scenarios; this second study allows for deeper analysis and, if needed, quantification of a risk level.

8.3. Model-driven PHA

The term *model-driven risk analysis* refers to a risk analysis approach based on an explicit model of the system under consideration and identified risk scenarios. The analysis process is represented using a structured model rather than an informal table containing textual information with no particular semantics. The model may be used as a data model for computer aided tools. Such an approach presents the following advantages:

– The model allows us to maintain greater consistency between different scenarios (rows in the table), as elements used in several places refer to a single element of the model.

– For the same reasons, modifications are propagated from one part to another part of the model.

– Transformations from one representation to another, from a table to a graphical representation and vice versa, for example, are easy to implement.

– The formalization of the representation makes it possible to build and share knowledge, as the information is organized and may serve as a basis for the construction of "data base for risk analysis".

System modeling constitutes a preliminary stage in the implementation of a model-driven approach. In the following stage, the model serves as a basis for the construction of scenarios in the PHA table:

– Each undesirable event is an event linked to a system, a sub-system, a structural entity or a function.

– Each hazardous situation is defined in giving a sub-system, entity or activity and an event related to this element. The event may be (section 7.9):

- a fault or failure;

- a parameter deviation;

- another undesirable event;

- degradation;

– an abnormal event of another type that does not fit into the above categories.

– If the causes of the hazardous situation need to be identified, the same typology is used.

– Consequences are described in the same way as causes, in giving a subsystem, entity or activity and an event related to this element, which may be:

- a fault or failure;

- a parameter deviation;

- another undesirable event;

- degradation;

- an abnormal event of another type that does not fit into the above categories.

Barriers are associated with events, for which they either prevent the occurrence of causes, the case in which they come before the event in the causality chain, or prevent the occurrence of consequences, the case in which they come after the event.

Following this approach, we implicitly build an event graph describing undesirable system behaviors. This graph allows us to represent PHA situations in a consistent manner and provides a basis for the construction of a bow-tie diagram or fault tree. It also allows us to create a link with FMEA- or HAZOP-type analyses, as we will see later. Software packages may be used to facilitate handling (Appendix 8).

8.4. Variations of PHA

8.4.1. *Different forms of results tables*

As we have already stated, PHA is not standardized, and a number of different variations exist. In this section, we present a selection of the different types of results tables used in practice. Section 7.4.2 will present a specific form used in the chemical industry.

Figure 8.9 shows entities linked to an undesirable event, in addition to the system and causes of the hazardous situation. The undesirable event is referred to as an accident. The table only shows the risk level, with no probability or severity details. Finally, it specifies proposed measures using three categories: technical, human and organizational.

Preliminary hazard analysis									
System	Product or equipment	Hazardous situation	Event causing the situation	Potential undesirable accident	Consequences	Risk level	Preventive measures		
							Technical	Organization	Human

Figure 8.9. *Another PHA table format*

Figure 8.10 presents a simplified form that is suited to use in evaluating occupational risks. It simply shows the undesirable event, the hazard family and possible dangers.

Preliminary hazard analysis							
Unit	Activity or Entity	Hazard family	Undesirable event	Damages	L	P	Preventive measures
Kitchen	Meal preparation	Burns	Accidental hot water spillage	Minor burns			Recommend caution

Figure 8.10. *Simplified format*

Figure 8.11 shows a format proposed by the INERIS (*Institut national de l'environnement industriel et des risques*) [INE 06], suitable for use in studying hazards in industrial installations. Consequences are expressed in the form of hazardous phenomena. The cause of an undesirable event is referred to as the drift, corresponding to the hazardous situation. The cause of the drift is also included in the table, allowing us to provide an additional level of causality.

8.4.2. *PHA in the chemical industry*

The *Union des industries chimique* (UIC), the French Union of Chemical Industries, provides a PHA method specifically developed for the chemical industry. The implementation of PHA in these cases follows a specific method, as activities in this domain generally involve handling a number of potentially hazardous products. Therefore, users create specific sheets describing products and procedures. Product data sheets specify all of the physicochemical properties of the product in question, notably those linked to safety (toxicity, flammability, etc.). Subsequently, these product data sheets and descriptions of reactions are used in conjunction with procedure descriptions, containing the conditions of operation and, where available, studies on the hazardous aspects of the procedure, to build the following tables:

– Material: an example is shown in Figure 8.13.

– Storage: this table provides an inventory of stored products, specifying quantities and physical states.

– Reactions: this table gives a detailed description of reactions, their chemical and physical characteristics, the conditions of occurrence, etc.

– Separations–treatments: in the same way, this table describes separation and treatment.

– Equipment: this table describes containers.

– Mode of operation and environment: this table describes the mode of operation.

N°	Causes (of deviation)	Deviation	Central undesirable event	Hazardous phenomenon	Frequency of cause	Intensity (from 1 to 4)	Severity for populations	Severity for environment	Safety barriers					NC
									Title	Cause	Deviation	CUE	DP	

Figure 8.11. *INERIS format for hazard studies*

Hazards are coded using the codes given in Figure 8.12. As a whole, this analysis process allows us to identify hazards. It is then possible to continue to a more detailed analysis (Figure 8.11).

F		Fire, i.e. combustion of condensed matter (flammable liquid leak, dust layers, etc.)
E		Explosion of purely physical origin (passage from a high-pressure circuit into a reservoir, sudden vaporization of a liquid, etc.)
EG		Explosion of gas or vapors, corresponding to an explosive process, generally a combustive – fuel mixture
EP		Explosion of powder in suspension: explosive process similar to that described above
ET		Thermic explosion resulting mainly from an uncontrolled chemical reaction (temperature runaway, gas formation, exothermic decomposition)
D		Detonation: explosive process occurring in specific conditions, either in gaseous phase with certain pure (acetylene) or mixed (hydrogen-air) compounds, or in condensed phase with compounds (peroxides) or mixtures (nitric-oxidizable derivatives) with high energy content (see CHETAH index)
T		Toxicity-agressivity, covering risks linked to immediate effects (acute toxicity, agressiviry) or delayed effects (chronic toxicity, work-related illness, carcinogenic substances)

Figure 8.12. *Hazard coding*

Procedure		PHA			Page	
Section					Date	
MATERIALS						
PRODUCT	PHYSICAL STATE	CURRENT QUANTITY	HAZARD	RISK AND PREVENTION	DETAILS NOTE No.	
REAGENTS						
PRODUCTS						
CATALYSTS						
SOLVENTS						
INTERME-DIARIES						
EFFLUENTS WASTE						

Figure 8.13. *PHA materials table*

8.5. Examples of application

8.5.1. *Desk lamp*

Let us consider a simple textbook example used to illustrate the implementation of the method. This example concerns a desk lamp and its environment. For our purposes, we will consider that the lamp may generate the following damages:

– Minor injuries in case of shock or entanglement with a user.

– Electric shock, when the power cable is bared as a result of repeated unplugging by pulling on the cable, and in cases where the differential circuit breaker does not work correctly.

– Burns, if a user attempts to change the light bulb without waiting for it to cool.

– Fire, in cases of prolonged use, resulting from poor ventilation due to unsuitable placement of the lamp.

The lamp-environment model is given in section 7.7. It is made up of three systems:

– the lamp;

– the user;

– the environment surrounding these elements.

In this example, we will use the severity and probability scales defined in section 6.5 for occupational risks.

The undesirable events include:

– electric shock, which may occur if the cable is bared and the circuit breaker fails to operate correctly. Consequences of this event include death, and are, therefore, assessed at level S1;

– overheating, if the lamp does not have correct ventilation. The consequence of this event is fire, with a significant severity level, assessed at S1;

– contact with a hot light bulb, if a user attempts to change the bulb without following the usage instructions and fails to let the bulb cool. The consequence of this event might be minor burns, assessed at S5 on the scale;

– a bump or trapped fingers, for example, leading to minor bruising, assessed at S5.

These events are shown in a PHA table (Figure 8.14), which gives a general overview. The likelihood is assessed in a qualitative manner based on feedback.

	Preliminary hazard analysis				
System	Hazardous situation	Undesirable event	Target and effects	P	S
Lamp	Changing light bulb	Contact with hot light bulb	Injury through burning	C	S5
Lamp	Bared wire Failure of electrical protection	Electric shock	Injury, death	E	S1
Lamp	Sudden or abrupt movements	Physical shock	Minor injuries	D	S5
Lamp	Poor ventilation	Overheating	Fire	E	S1

Figure 8.14. *PHA of the desk lamp*

8.5.2. *Chemical reactor*

This system is described in Appendix 7. The reactor is used to carry out an exothermic chemical reaction between two reagents A and B to produce a product C. Products B and C are toxic. The probability and severity scales for industrial risks described in section 6.6.2 will be used in carrying out this analysis.

The undesirable events are, first, linked to a loss of control of the reaction:

– The first undesirable event is an increase in pressure, which, when occurring at the same time as a blockage of the safety valve, may lead the reactor to explosion, releasing large quantities of the toxic product, posing a threat to the surrounding population.

– This increase in pressure may be the result of:

- a problem related to loading, which may be excessive;

- a mixing problem, preventing homogeneous cooling;

- a cooling problem due to the double jacket loss of efficiency.

This second undesirable event is one of the causes of the first UE.

Other undesirable events identified for this installation are:

– leaks from pipes, potentially leading to intoxication of personnel;

– contact with voltage-carrying components during maintenance activities.

The severity of these different events is assessed based on their potential consequences. The probability is assessed in a qualitative manner, using an approach based on feedback and the mixed scale defined for industrial risks (section 6.6.2).

Preliminary hazard analysis						
System	Hazardous situation	Undesirable event	Target and effects	P	S
Reactor system	Excess pressure Blocked valve	Explosion	Destruction Physical damage Pollution	C	S1
Reactor system	Wear and tear to joints and valves	Leak (joints, valves, safety valves)	Intoxication of personnel	D	S2
Reactor system	Poor labeling of live elements	Contact with live elements	Electric shock to maintenance personnel	D	S2
Cooling system	Poor temperature regulation No cold water supply Failure of P33040 pump	Insufficient cooling	Excess pressure	C	S1

Figure 8.15. *PHA for the chemical reactor*

8.5.3. *Automobile repair garage*

This example deals with an automobile repair garage made up of a workshop, an area used for storing spare parts and materials, an administration area and an outdoor parking area. In a structure of this type, a simplified form of PHA may be used in evaluating occupational risks in order to create a inventory of occupational risk. The PHA process is limited to identifying the undesirable event; the target is not specified, and is generally considered to be the operator. The hazardous situation is not described in detail.

The garage is divided into four work units: workshop, administration, storage and parking lot. These units are represented by a system. Each unit is analyzed using the occupational risk checklist presented in Chapter 3. Figure 8.16 shows an extract from the PHA table concerning the garage. Each entity is analyzed in relation to the dangers it presents. The undesirable event describes the risk, which is then evaluated using the scale mentioned in section 6.5; preventive measures are also proposed.

Preliminary Hazard Analysis 173

S	Entity/Activity	Hazard family	Undesirable event	L	S	Means of prevention
Workshop	Lifting platform	PP5 Mechanical	Crushing by the car, shearing during installation of vehicle	C	S1	Platform commands fixed to the column and controlled by platform operator; emergency stop provision
	Lifting platform	T1 Fall from a height	Fall from the platform or from a ladder while lifting a vehicle for repair	B	S3	Safety instructions
	Lifting platform	T5 Manual handling	Physical shocks leading to minor industry	B	S2	Use of workshop cranes for motors. Use of load lifters for gearboxes
	Vacuum station	PP2 Electrical	Contact with live aspects leading to electric shock or electrocution	C	S1	30 mA circuit breaker, earthed electrical installation
	Air conditioning recycler	PP5 Mechanical	Flex under pressure, risk of projection of the flex	C	S4	Training before using equipment
	Air conditioning recycler	PP6 Pressure Equipment	Injury due to the use of 134A under pressure	C	S3	Secured rapid connector
	Air conditioning recycler	PP2 Electrical	Contact with live aspects leading to electric shock or electrocution	D	S1	30 mA circuit breaker, earthed electrical installation
	Use of jack	ACT5 Posture	Repetitive Strain Injury (RSI)	D	S4	
	Use of jack	T5 Manual handling	Injury, Repetitive Strain Injury (RSI)	B	S3	Weight limits for lifting. Use of caddies to move oil containers. Use of wheeled containers for emptying
	Use of jack	PP5 Mechanical	Crushing under car	C	S2	Jack used with safety stand
	Tire changer	PP5 Mechanical	Projection of wheel rim and/or tire	C	S3	Control pedal on machine
	Damper compressor	PP5 Mechanical	Projection of damper during compression	C	S1	Mobile guide grid, remains mobile when machine is in use
	Grinder	PP5 Mechanical	Projection, catching or burns	C	S2	One grinder with a protective cover, the other without, use of goggles
	Degreasing washer	T3 Chemical	Corrosive products	D	S3	Use of PPE (goggles, gloves)

Figure 8.16. *Extract from the PHA of a garage*

8.5.4. *Medication circuit*

The medication circuit is a circuit that involves a number of different aspects of health care establishments. Each stage in the circuit is a source of errors that may affect patient safety. In France, a legal order was published on April 6, 2011 with the aim of making mandatory risk management practices for this circuit. This process includes a PHA phase.

The creation of a full model of the medication circuit for a hospital is a huge task. For the purposes of PHA, we will represent the installation using six systems:

– S1: prescription;

– S2: dispensation, preparation and supply;

– S3: transportation;

– S4: holding and storage;

– S5: administration;

– S6: patient information and monitoring.

Iatrogenic (medicine-related) diseases are defined by a set of potential or observed consequences harmful to human health resulting from medical intervention, healthcare or the use of a healthcare product. An iatrogenic medication error (IME) is defined as an avoidable deviation from what should have taken place in the drug-based treatment of a patient, which may cause a risk or an undesirable event for the patient.

Each system is analyzed using the appropriate checklist in order to identify undesirable events that may lead to medication errors. An extract from the PHA table is shown in Figure 8.17.

System	Undesirable event	Hazardous situation	Consequences	L	S	A	Possible means of prevention
Prescription	Prescription includes medicines which interact or have contraindications	Prescription created using all available information	Secondary effects for patient	D	S2	Unacceptable	Screening of drug given in relation to patient symptoms
Prescription	Prescription includes medicines which interact or have contraindications	Prescription created without respecting the best use contract	Secondary effects for patient	E	S2	Unacceptable	Avoid prescriptions given by individuals other than certified professionals
Prescription	Prescription includes medicines which interact or have contraindications	No system exists to monitor interactions and contraindications in healthcare facilities	Secondary effects for patient	E	S2	Unacceptable	Implement monitoring system
Prescription	Prescription of an unavailable drug	Drug prescribed using specialist name and not by INN	Delay in patient treatment	D	S3	Unacceptable	Popularize names
Prescription	Prescription of an unavailable drug	Drug not listed in therapeutic register	Delay in patient treatment	B	S3	Unacceptable	Monitor drug stocks

Figure 8.17. *Extract from the medication circuit PHA*

8.6. Summary

PHA is one of the most widespread risk analysis methods. It may take a variety of different forms suited to different domains of activity. PHA is one of the most widely used methods for use in identification and qualitative or semi-qualitative analysis.

Hazard	Undesirable phenomenon
Heat	Burns
	Overheating
	Inflammation
	Increased evaporation
	Non-intentional start to reaction
	Increase in pressure
	Loss of mechanical properties
	Heat-related dysfunction
	Physiological problems for humans
Cold	Degradation due to freezing
	Loss of mechanical properties
	Cold-related dysfunction
	Physiological problems for humans
Light	Glare
	Lack of light
Non-ionizing radiation	UV radiation
	Laser radiation
	Microwave
	High frequencies
	IRM
Ionizing radiation	Alpha
	Beta
	Gamma
	Neutron
	X rays
Mechanical	Ejection or projection (explosion, bursting)
	Injury due to cutting or pointed aspects
	Crushing (objects, persons)
	Punching
	Catching by a rotating element
	Collapse
	Instability
	Vibrations
Electrical	Contact with live element
	Overheating, start of fire
	Elecrical arc due to HT
	Sparks due to static electricity
	Power cut

Table 8.1. *List of hazards and associated undesirable events*

Hazard	Undesirable phenomenon
Kinetic energy, gravity	Sudden movement of machines
	Impact of moving elements
	Loss of control of moving elements
	Objects or persons falling from a height
	Objects overturned, persons falling from standing position
	Projection of debris, missile effect
	Wave effect in liquid ("sloshing" in partially filled containers)
Fire	Presence of fuel
	Presence of combustibles
	Flammable matter
	Source of ignition
Explosion	Presence of explosive atmosphere
	Presence of explosive liquid
	Presence of explosive powder
	Presence of trigger (heat, sparks, shock, lightning, electricity, etc.)
	Shockwave
	Projection of elements
Hazardous materials	Fire
	Explosives
	Flammable materials
	Combustibles
	Pressurized gas
	Corrosives
	Toxic or neurotoxic material
	Irritant, sensitizing or narcotic materials
	Sensitizing, mutagenic, carcinogenic or reprotoxic materials
	Environmental hazard
	Reacts with air, water or other
	Cryogenic liquid or product
	Pathogenic (bacteria, HIV, other virus)
	Allergenic
	Source of smell
	Loss of containment
Pneumatic & hydraulic pressure	Excess pressure
	Rupture in container or piping
	Explosion and projection of elements
	Implosion
	Whip effect due to tube rupture
	Water hammer, cavitation
	Safety valve failure
	Backwash

Table 8.1. *(Continued) List of hazards and associated undesirable events*

Preliminary Hazard Analysis

Hazard	Undesirable phenomenon
Ergonomics	Error due to: - lack of emergency stop button - excessive fatigue - inadequate user interface on machines - workstation design fault - inadequate operating procedure - accessibility problem - lighting problem
Working conditions	Error due to: - stress - poor training - inappropriate instructions - non-respect of indications
Operator	Unsuitable action
	Premature action
	Late action
	Actions carried out in wrong order
	Lack of action
	Poor understanding of information
	Lack of information
	Poor transmission of information or instructions
Utilities	Loss of electricity
	Loss of vapor
	Loss of heating/cooling
	Loss of ventilation
	Loss of conditioned air
	Loss of compressed air
	Loss of lubrification
	Loss of fuel feed
	Loss of means of evacuating used gas
	Loss of data network
Electronic control system	Poor cleaning
	Power supply problem
	Interference due to unsuitable electromagnetic compatibility
	Humidity
	Electrical fault
	Software fault
	Lightning strike
	Earthing fault
	Sudden or abrupt actions

Table 8.1. *(Continued)List of hazards and associated undesirable events*

178   Risk Analysis

Hazard	Undesirable phenomenon
Major external phenomena	Flooding
	Fire
	Storm
	Freezing
	Earthquake
	Snow fall
	Blackout (electrical)
	Magnetic storm
	Pandemic
	Loss of internet connection
	Loss of telephone network or GSM
	Road traffic blocked
Operations	Repetitive Strain Injury (RSI)
	Manual handling of heavy loads
	Handling of objects which may cause injury
	Problems during:
	- transportation
	- settings
	- verification
	- delivery/reception
	- startup
	- start
	- emergency start
	- stop
	- emergency stop
	- testing
	- diagnostics
	- maintenance
Shared causes	Loss of utilities
	Major external phenomena
	External temperature
	Humidity
	Calibration error
	Maintenance error
	Pest control issues

Table 8.1. *(Continued) List of hazards and associated undesirable events*

Chapter 9

Failure Mode and Effects Analysis

9.1. Introduction

The failure mode and effects analysis (FMEA) is, just like the preliminary hazard analysis (PHA), an *a priori* risk analysis method, that is to say a method used to analyze possible risks before any accident occurrence. However, contrary to the PHA method, its objective is not to find the undesirable events that might bring about damage, but to examine all of the failure modes of a system, to analyze their effects and to search for their causes. Moreover, when we assess the criticality of each of these modes, we talk of a failure mode, effects and critically analysis (FMECA) method. Oftentimes, the term FMEA is used to broadly designate the method.

There are several types of effect of a failure mode, and most of them do not impose damage upon a target, but rather an interruption to the system's functioning or the lack of capacity to fulfill the needs it has been designed for. Consequently, the FMEA method is used in several fields besides risk analysis, and it is particularly used in the context of quality management (ISO 9000 standard) or for product design. We sometimes find names for the method that have been adapted to the object of study, such as:

– the Design FMEA, which analyzes the possible failures of the product, due to its design, its development, its manufacturing or its use;

– the Process FMEA or Procedures FMEA, which is used for studying the potential defect of a new product or an old product, a defect that comes up during the manufacturing process;

– the Production Means FMEA which is used to analyze potential failures of production means (factories, machines, and tools used to produce);

– the Service FMEA, which is used for analyzing the failures of the service production systems;

– the Organization FMECA, whose objective is to analyze the failures of an organization.

Within the process of risk assessment, the FMEA method is generally used after a phase of risk identification, and it allows us to make the connection between the failures of the different elements and the undesirable events that can take place (Figure 9.1).

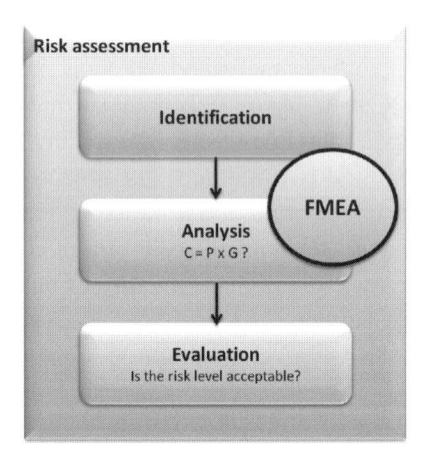

Figure 9.1. *FMEA in the risk assessment process*

This method was developed by the American Department of Defense at the end of the 1940s (MIL-STD-1629A), and then in the 1960s was used to study the reliability/viability of spatial missions. In the 1970s, it was one of the tools used to study the reliability of the design of nuclear plants. During this same period, it was used in the automotive industry, in particular by Ford where it was introduced in the Q101 quality standard in 1986.

Today, it is defined by the IEC standard 60812, *analysis techniques for system reliability – procedure for failure mode and effects analysis (FMEA)*, published in 2006, which describes the modes of application for the reliability study, and by standard SAE J-1739 of the automotive industry, *potential failure mode and effects analysis in design (Design FMEA), potential failure mode and effects analysis in manufacturing and assembly processes (Process FMEA),* which describes its use in the design and assembly processes.

9.2. Key concepts

9.2.1. *Basic definitions*

The basic terms used within the FMEA method, and defined by the standard, are the following:

– *Device:* any element, component, subsystem, functional unity, equipment or system that can be considered on an individual basis. A device can be made of hardware, software or both at the same time, and can also include the workforce in certain cases. A fixed set of devices, for example a population or a sample, can itself be considered as a device.

– *Failure:* ceasing of an entity's ability to accomplish a required function. A classification of the possible failures is discussed in section 3.4.

– *Fault:* state of an entity that is unable to carry out a required function, not included the inability during preventive maintenance or other planned actions, or due to a lack of external means.

– *Failure mode:* the manner by which a failure is observed.

In all of these definitions, the notion of a *failure mode* is what is at the core of the FMEA method. It is the way in which the failure manifests itself on the expected behavior. Figure 9.2 shows examples of failure modes. We can identify generic modes (Figure 9.3), but these modes must be adapted to the field of activity and at the level of detail of the analysis. A more detailed list of generic modes (Figure 9.4) given in version 1985 of the standard X60-510 may be used as a reference. Since it is relatively specialized, it does not show up in the current version of the standard.

9.2.2. *Causes of failure*

The *cause of a failure* (Figure 9.5) may be defined as an event related to the design, manufacturing or use of the device, which can bring about a failure. Within an FMEA analysis, it can be useful to identify the most probable causes for each potential failure mode. Identifying these causes is not always necessary: this can be reserved for the modes whose effect has a significant impact. These causes allow for a better assessment of the probability of the mode and enable us to foresee prevention measures, by acting at the level of these causes.

N°	Element	Examples of failure modes
1	Switch	Open, closed, no contact
2	Valve	Open, partially open, closed, partially closed, leaking
3	Spring	Crushed, broken, blocked in extension
4	Relay	Open, closed, false contact, short-circuited coil
5	Pump	Does not pump anymore, flow is too weak, flow fluctuates
6	Operator	Wrong operation on the right element Wrong operation on the wrong element Action carried out too early Action carried out too late Action not carried out Wrong action

Figure 9.2. *Failure modes examples for various devices*

Generic failure modes | Behaviour expected / real

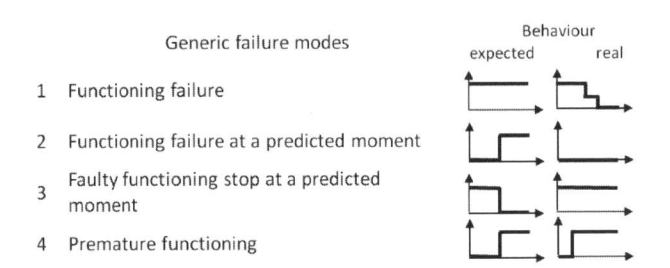

1 Functioning failure
2 Functioning failure at a predicted moment
3 Faulty functioning stop at a predicted moment
4 Premature functioning

Figure 9.3. *Generic failures modes*

We may note that one mode can have several independent causes, and that conversely, a cause can be at the origin of several modes, which is the case of common cause failure modes. We will detail this further on.

Furthermore, a cause can be internal or external to the examined element. For example, an electrical device may have a breakdown on an internal circuit, or can be perturbed by electric parasites. The causes can also come from other stages in the life of the system, such as the maintenance phase or the design phase. To find out the causes, the classification is proposed in Chapter 3.

1. Structural fault	18. Incorrectly started up
2. Physical blocking	19. Does not stop
3. Vibrations	20. Does not start
4. Does not stay in the right position	21. Does not switch over
5. Does not open	22. Prematuring functioning
6. Does not close	23. Functioning after the predicted period (delay)
7. Faulty in open position	24. Wrong input (increase)
8. Faulty in closed position	25. Wrong input (decrease)
9. Internal leakage	26. Wrong output (increase)
10. External leakage	27. Wrong output (decrease)
11. Surpasses the limit	28. Input loss
12. Is below the lower limit	29. Output loss
13. Overfunctioning	30. Short circuit (electric)
14. Intermittent functioning	31. Open circuit (electric)
15. Irregular functioning	32. Leakage (electric)
16. Wrong indication	33. Other exceptional faulty conditions following the system characteristics, the functioning conditions and operational constraints

Figure 9.4. *List of failure modes (standard 60812:1985)*

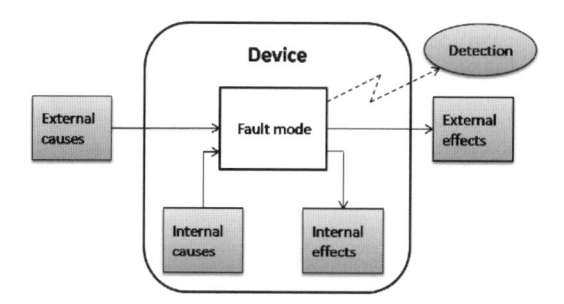

Figure 9.5. *Failure mode*

9.2.3. *The effects of a failure*

The effect of a failure is defined as the consequence of the failure mode upon the behavior or the state of the device. The effects can be seen on:

– the availability of the production means;

– the quality of the product;

– the costs;

– the safety of people and goods;

– the environment.

We distinguish between the *local effects*, which are the consequences upon the outlet of the device which is subject to the failure, and the *overall or final effects* at the level of the system in its entirety, which are caused by the local effects.

In order for an effect to occur, it may require several failures to take place simultaneously. This is, in particular, true for overall effects.

EXAMPLE 9.1.– Let us consider a electrical power system with an backup battery (Figure 5.15). The failure of the backup battery results in the unavailability of the entire system if the main battery fails.

9.2.4. *Frequency or probability of a failure*

The *frequency* or *probability* of a failure enables us to characterize its likelihood. It is useful to mention the period of time used to specify this frequency. It can be assessed either qualitatively or quantitatively. In those cases where the failure mode relates to a relatively simple component, it can be estimated from the failure rate by taking into account the operating conditions.

The standard EN60812 proposes the scale shown in Figure 9.6, as well as a scale on 10 levels used in the automotive industry. A scale suited to the problem must be defined by following the general principles described in Chapter 6.

Level	Description	Probability
1 or E	Unlikely	$p<0.001$
2 or D	Remote	$0.001<p<0.01$
3 or C	Occasional	$0.01<p<0.1$
4 or B	Probable	$0.1<p<0.2$
5 or A	Frequent	$0.2<p\leq1$

Figure 9.6. *Example of a frequency classification*

9.2.5. *The severity of a failure*

The *severity of a failure,* denoted by S, is defined by the standard as an assessment of the significance of the failure mode's effect on the operation of the device or on its environment. The severity of the effect of a failure mode depends on the limits defined for the system analyzed and is measured by examining the overall or final effect. This notion corresponds to the *severity* of the risk.

As with the other methods of risk analysis, the severity can be represented by a symbolic or numerical value. The standard proposes several classifications. We will present a classification with 4 levels (Figure 9.7). There is another classification with 10 levels. The classification must be tailored for each particular case.

Level	Description	Meaning
4	Catastrophic	Serious damage to the system Human injury
3	Critical	Considerable damage to the system No serious threat of injury or other life-threatening conditions
2	Marginal	Degradation of system performance without significant damage to the system or any threats of fatal injury
1	Insignificant	Potential degradation of the system's functions but no threats of life-threatening injury

Figure 9.7. *Example of severity classification*

9.2.6. *Detection of a failure*

A failure mode can be more or less easily *detected*. The easier we can detect it, the easier it will be to take the necessary measures to limit its effects. For example, if the failure threatens safety, we will be able to implement a safety system when a failure is detected. In the case of a failure that has an effect on the production or on its level of quality, the detection allows us to take appropriate measures by reorganizing the production capabilities.

In an FMEA analysis, we can therefore list the means for detecting each failure mode. These detection means can be seen as a protection barrier against the effects of the failure. They can, for example, take the form of sensors, monitoring via statistic control or monitoring via an operator. It is worth noting that several failures can manifest themselves identically and that the detection is not always obvious.

The detection means are characterized by a probability of non-detection and this parameter is included in the level of criticality. It can be either qualitative or quantitative. It is larger as the mode is less detectable. The standard proposes the scale shown in Figure 9.8.

Level	Detection	Meaning
4	Non-detectable	Non-controlled mode
3	Weak	Difficult detection
2	High	Detection carried out by sampling
1	Certain	Automatic detection

Figure 9.8. *Example of a detection level scale*

9.2.7. *Criticality of a failure and RPN*

The criticality of a failure is defined by the standard as the combination of the severity of its effects and its likelihood, which corresponds to the usual definition. To determine this, we can use a table with two entries, the probability or frequency and the severity. A certain level of criticality is defined for each cell in the table. For the scales of probability and severity defined above, the standard defines four levels: intolerable, dangerous, tolerable and negligible (Figure 9.9).

	E or 1	D or 2	C or 3	B or 4	A or 5
4	Tolerable	Undesirable	Intolerable	Intolerable	Intolerable
3	Tolerable	Undesirable	Undesirable	Intolerable	Intolerable
2	Negligible	Tolerable	Undesirable	Undesirable	Intolerable
1	Negligible	Negligible	Tolerable	Tolerable	Undesirable

Figure 9.9. *Risk matrix for defining the criticality*

In FMEA analyses, we often encounter an index called *risk priority number* (RPN), which characterizes the level of criticality, and which is defined as the product between the probability and the severity index:

$$RPN = P \cdot S$$

If the detection level is assessed, the RPN will include this parameter:

$$RPN = P.S.D$$

As we have seen in Chapter 6, the use of a risk matrix allows for more flexibility.

9.3. Implementation of the method

The FMEA method is an exhaustive method for analyzing the effect of all the failures of *all* the elements of a system. It successively considers the manner by which a failure is observed, and therefore assumes the implicit hypothesis that there are no *simultaneous failure modes* that manifest themselves at the same time. This hypothesis is important and is a limitation in considering the failures that have a common cause. We will return to this in the remainder of this chapter (section 9.5.1).

We can study the failures of the functions of a system, those of the components, or structural entities of a system, or a mixture between the two.

The *functional* approach is used;

– when the components cannot be identified, for example very early in the design process;

– or if the complexity of the system needs to structure the analysis by using functional blocks.

The *components or hardware* approach is used:

– if the components can be identified and they have easily identifiable features;

– if the analysis, in order to be quantified, needs to be based on reliability data depending on the physical component.

The *mixed* approach is also used quite often. It is built upon a functional analysis of a system in order to better understand it, complemented by the identification of the components used by the functional groups.

Whatever the type of approach chosen, the stages of an FMEA analysis are the following (Figure 9.10):

1) Analysis preparation: context definition, information gathering and observations.

2) Description and modeling of the installation or system.

3) Application of the FMEA procedure:

   i) failure mode identification;

   ii) effect analysis;

   iii) cause analysis;

   iv) probability assessment;

   v) severity assessment.

4) Review of the analysis and of the measures to be taken, drawing up the analysis report.

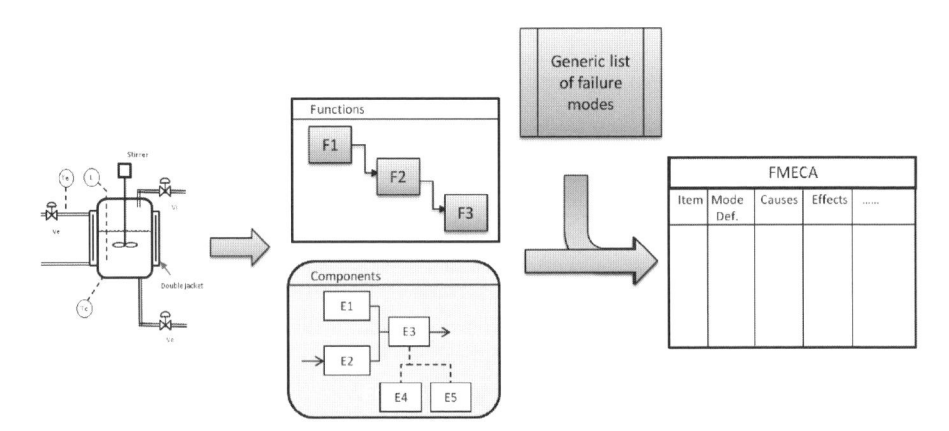

Figure 9.10. *Overview of the FMEA*

We will detail these stages in the following section.

REMARK 9.1.– In most cases, the effect of failures does not have any damaging consequences. Therefore, in order to increase the efficiency with which the FMEA method is implemented, it is preferable to:

– perform a PHA analysis or a hazard identification before implementing the FMEA method;

– try to limit the failure analysis to those leading to a potentially undesirable event.

Therefore, the FMEA method will be a means of analyzing in more detail the scenarios leading up to the undesirable events and the respective damages. It will be followed by the construction of a detailed failure tree leading up to that dangerous event.

9.3.1. *Analysis preparation*

This phase is part of the definition of the risk management context. It includes:

– defining the objectives of the analysis (safety, quality, etc.);

– planning the analysis;

– defining the limits of the system analyzed and the level of detail;

– creating the work group, ultimately involving experts for certain elements.

Next, it is necessary to define an adequate scale of the severity and frequency indices; this scale will be either qualitative or mixed (if certain probabilities are assessed on the basis of failure rates) or will be used in failure trees in the following stage. We will next define how the criticality is calculated: with a risk matrix or using the RPN index.

Finally, we can specify if the detection means are analyzed systematically and, if they are indeed assessed, we can also specify an adequate scale.

9.3.2. *System modeling*

Between defining the context and starting the FMEA analysis, it is necessary to model the system. The approach used most often is a functional analysis (section 7.3), whose aim is to:

– structure and modularize the analysis;

– understand how and why the entity is working;

– provide a representation of the system that will facilitate the systematic research of failure modes.

This functional analysis can be completed or replaced by a structural analysis in the case of a component-based FMEA. In this modeling stage, we can identify redundancies. The model may given in a tabular form (Figure 9.11).

System	Function / Entity	Description	Inputs	Outputs
(1)	(2)	(3)	(4)	(5)

Figure 9.11. *Example of table for model description*

9.3.3. *Application of the analysis procedure*

Once the installation has been modeled, we must identify redundancies to examine the different elements by following the process described in Figure 9.12. The result of the analysis of each mode allows us to fill in the table according to the format shown in Figure 9.13. If the means available for carrying out the analysis are limited, it may be useful to choose the 20% of the equipment that generates 80% of the criticality.

To choose the functions that generate criticality, one solution would be to identify the events undesirable at the level of the system. We will then work on the failure modes that lead up to these undesirable events.

9.3.3.1. *Identifying the failure modes*

For each selected function or entity, we look for its failure modes:

1) The first failure mode will generally be of the type "total loss of the function or component". It may be useful to distinguish the states that the system finds itself in after the failure, because the effects will be different. For example, a switch may be blocked either on or off, which has very different consequences.

2) If necessary, we will examine the different possibilities of partial loss of the function or the component, whether it is in terms of amplitude of the function's specifications, such as "partial implementation of the function", or in terms of the functioning in relation to time, such as "untimely functioning", "delayed start", "premature stop" and "too long/too short functioning period".

3) It can also sometimes be useful to examine the failure cases due to a design error: "specification of the inadequate function or component".

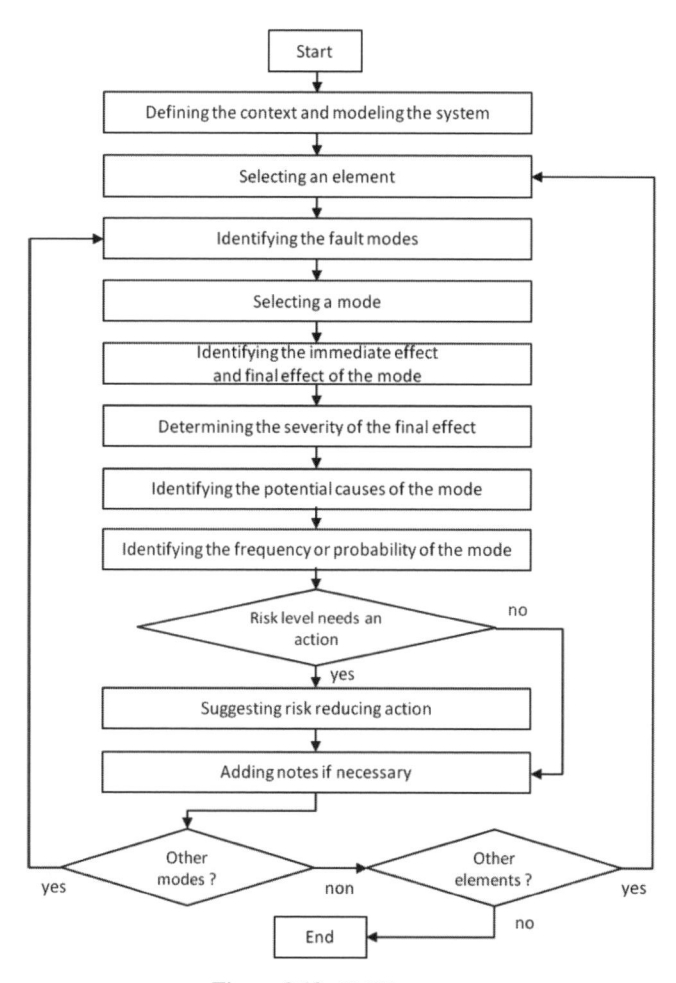

Figure 9.12. *FMEA process*

192   Risk Analysis

System	Function Entity	Failure modes	Possible causes	Possible effects	S	P	Detection means	Existing risk prevention action	Notes
(1)	(2)	(3)	(4)	(5)	(6)	(7)	(8)	(9)	(10)

Figure 9.13. *Example of FMEA results table*

For each mode identified, we create a new row in the table and we add the mode in column 3, while columns 1 and 2 allow us to specify the system and the function or entity under consideration.

EXAMPLE 9.2.– Let us take, for example, the function "cooling by circulating cold water". We will be able, for instance, to identify the following modes:

1) Does not refrigerate (we understand, implicitly, that it does not refrigerate due to the breakdown of an element).

2) Refrigerates insufficiently.

3) Does not refrigerate upon prolonged use.

4) Does not refrigerate if the outside temperature is too high (faulty design).

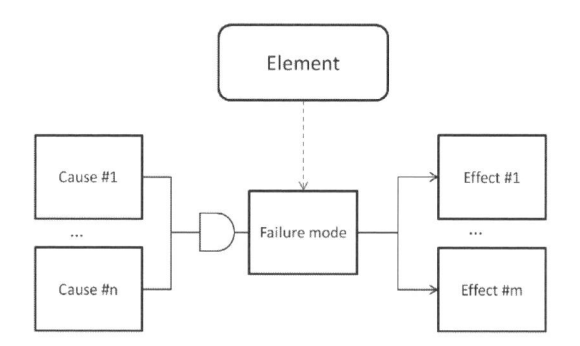

Figure 9.14. *Connections between failure mode, causes and effects*

9.3.3.2. *Effects analysis*

For each identified mode, we look for the immediate effect, that is the effect upon the nearest element, as well as for the global effect at the level of the installation. These effects are described in column 5. We can even add a column 5' in order to differentiate between the local effects and the global ones effects. The distinction is not always easy. The use of a model-based approach, however, helps overcome this difficulty (section 9.4).

A further ambiguity comes from the fact that the FMEA method is based on the assumption that there cannot be simultaneous modes. Consequently, the effect obtained presupposes that the set of other functions and entities are functioning correctly. In certain cases, this is not satisfactory and we end up adding effects under conditions.

EXAMPLE 9.3.– Let us take once again the example of a lamp, but this time adding a backup battery. The effect of a failure in power supply will either be non-existing if the battery functions well, or a loss of function if the battery fails. Not taking the latter case into consideration may pose a problem. In this case, we will also see that a model-based approach may help to clarify things.

9.3.3.3. *Causes analysis*

Once we have identified the effects, we then become concerned with the causes of the failure modes, which are described in column 4. Generally speaking, these can be failures of other functions of the components, internal faults or damages generated by a dangerous phenomenon. The latter can either be external to the installation or come from the installation itself. A cause can be external from a geographical point of view, such as an external failure connected to a loss in the power supply or an external undesirable event, such as a fire. The failure can also be external from a temporal point of view, such as a failure in a different phase of the system's lifecycle, a design error or a maintenance error. A model-based approach allows us to structure the causes analysis.

9.3.3.4. *Assessing the probability*

The probability (or frequency) of the occurrence of a failure mode can be determined either directly or indirectly, by examining the causes of the failure mode. If the causes are available, the probability of a mode can be calculated using the probability of its causes, and we will only take into account the probabilities of the modes considered to be primary, that is the modes for which the causes are not yet identified.

9.3.3.5. *Determining the severity*

The severity of a mode is defined depending on the significance and the quantity of damage that it brings about. Oftentimes, a failure mode only generates indirect damage through the undesirablephenomena that are included in its global effects. We will therefore rely on these in order to assess the severity.

REMARK 9.2.– However, this approach can turn out to be incorrect in situations where a conjunction of events is necessary in order for the effect to take place. This is, for example, the case when the effect is the result of the conjunction between the failure of two elements, or when it is the result of the failure of one element and the failure of prevention barriers (i.e. multiple failures). Under these conditions, considering the severity of the effects of only one of the modes, while assuming that the others are not active, will lead to an underestimation of the severity; while taking into account the severity together with the other modes considered as being active will lead to an overestimation of the criticality, which is a product of the probability and the severity.

The solution lies in exhaustively describing the risk scenarios and assessing the different values of (p,s) for each consequence (section 6.3).

9.3.3.6. *Detection*

In column 8 (detection), we indicate if it is possible to detect the failure once it takes place, and what will be the means used to detect it. In certain cases, this detection regards the causes of the failure mode, and not the mode itself. This column is not always used.

EXAMPLE 9.4.– Let us take the example of the desk lamp. For the function "light up", we can easily detect this failure mode by means of visual observation. We can also try to detect one of the causes, for instance a broken light bulb, by examining it or testing it visually.

9.3.3.7. *Actions to be taken*

When the level of criticality of the examined mode is judged to be unacceptable, we need to propose a certain number of correcting actions. These can either act on the probability or on the severity, and allow us to obtain a new value of criticality. These new values are added in the table by filling in the corresponding columns.

9.3.4. *Review of the analysis and the measures to be taken*

The FMEA analysis enables us to highlight the critical failure modes of a system. A sum-up of the effects at the level of the system will be proposed in the final analysis report. For each effect, we will list the set of modes that lead to it and assess the probability of obtaining that mode.

Moreover, the report also draws up the list of the measures to be taken. In the simplest cases, the actions can be proposed directly. In the most complex cases, such as the case of failures that entail significant damage, or for the systems presenting redundancies or failure modes with a common cause, a more precise assessment must be carried out. We generally use a representation via a failure tree or a bow-tie diagram. This same representation will be used for assessing the level of efficiency of the prevention barriers.

Once the actions have been determined and prioritized, they must be implemented within a suitable process.

9.4. Model-based analysis

Just like PHA, the model-based analysis consists of describing the system with the help of a suitable model, and carrying out the FMEA analysis like the construction of a part in the dysfunction model. This is represented via an event graph that is enhanced by the FMEA analysis. The main advantage is to provide a complete representation of the connections between the cause and the effect that allow us to extract the FMEA table, to evaluate the probabilities and gravities of the basic (or terminal) events information, and to avoid the ambiguities regarding local and global effects, or the multiple failure modes.

In practice, the first stage consists of building a model of the installation, which will most often be a structural–functional model, in order to be able to analyze the effects of the failures of the functions as well as the components. The model can be represented in a tabular form (Figure 9.11). The description of the inputs and outputs of each element is optional. If it exists, it will be used to guide the FMEA analysis.

Applying the FMEA procedure should be carried out as previously, using this model, in order to identify failure modes. On the contrary, the effects are expressed by following the typology described in section 7.9, in which each event is either:

– an undesirable event connected to a system, or to another function or entity;

– a failure mode connected to a system, or to another function or entity;

– a deviation of a variable; or

– an event that characterizes a degradation.

With regard to the causes, they are expressed in the same way:

– either an undesirable event connected to a system or to another function or entity;

– as a failure mode connected to a system, or to another function or entity;

– as a deviation of a variable attached to a function or to a hardware entity;

– an event connected to a normal use of the system, which is a factor triggering the occurrence of the mode; or

– an event representing a combination of the above events.

An AND-type logical operator may be specified between these events.

If the system model describes the connections between the different functions and entities, the causality connections between the failure modes will follow along these relations: for example, the failure causes of a function will include all or part of the failure modes of the entities used at the input and the effects will be failure modes of the output elements.

EXAMPLE 9.5.– If the function *light up* uses one *light bulb* and one *battery*, the causes of the functioning failure will be the failure of its inputs, that is the failures of the light bulb and the battery, respectively.

By following this formalism, which allows us to build a dysfunction graph at the same time as the FMEA table, the problems of the local or global effects no longer arise, since we are describing the causality chain. The redundancies can be described by adding an AND operator. The common causes appear naturally since the same event appears as a cause for the different modes. Finally, by using this graph, the probability and severity calculations can be automated using computational tools (Appendix 8).

9.5. Limitations of the FMEA

9.5.1. *Common cause failures*

In a system, a certain number of failures are not independent because they can have a *common cause*. Let us mention, for example:

– Utility loss: electricity, steam, compressed air, nitrogen, IT network, WI-FI.

– A major external phenomenon: a fire, flood, storm, earthquake.

– A maintenance or calibration error, which will impact an entire equipment set.

– The action of harmful animals.

Taking into account common cause failures is in direct contradiction with the FMEA assumption, according to which only one failure can be considered at a given time. Common cause failures can, however, be considered informally in the table, by adding effects under certain conditions. To rigorously consider them, it is necessary to build a failure tree for the global effect under examination. By using a model-based approach, this tree will be drawn up at the same time as the FMECA table.

9.5.2. *Other difficulties*

Another limitation of the FMEA is the fact that it becomes tedious when we consider a complex system which has several functions and numerous components. This complexity can be increased if we have to consider several possible functioning modes and different lifecycle phases, such as operation and maintenance. In this case, it is preferable to begin by carrying out a PHA analysis and only analyzing the failure modes that are the causes of undesirable events or causes of other events that have already been examined (i.e. contributing to the occurrence of a undesirable event).

Another aspect that must be cautiously managed is the management of the different hierarchical levels described by the model. The analysis must be carried out in consequence. Let us take the example of a pressure limiting function, which needs a valve in order to work. Let us consider the undesirable event *pressure too high*. It is important to observe the hierarchy of the model throughout the analysis, that is to write down the failure of the function *limit pressure* as the effect of the valve failure, and the *pressure too high* as the effect of the failure of the function *limit the pressure*. Indeed, it is possible to put the undesirable event *pressure too high* as the effect of the valve failure and also as the effect of the failure of the function *limit the pressure*. The event *pressure too high* then appears as having two different causes. The analysis might then become incomprehensible.

9.6. Examples

9.6.1. *Desk lamp*

To illustrate the method, let us take the example of the desk lamp presented in section 7.7, whose preliminary hazard analysis is given in Figure 8.14. Each of the structural and functional elements of the lamp is reviewed in order to identify the failure modes, their effects and their causes. The result is shown in Figure 9.15. We notice that:

– the majority of the failures lead to a loss in the functionality of the lamp that no longer provides the required light;

– the failures leading to safety risks or health hazards for the user have as an effect one of the hazardous events that had been identified throughout the PHA. The FMEA analysis allows us, therefore, to analyze in more detail the causality chain leading up to these events.

We may also note that the IDEF0 model (Figure 7.14) allows us to structure the analysis, since the causes of the failure in the *light up* function are failure modes of its inputs, and the consequence of this failure is a failure of the output.

Failure Mode and Effects Analysis 199

System	Function or entity	Failure mode	Causes	Effects
Lamp	Producing light	Does not produce light	Light bulb does not light up; loss of electrical continuity of the cord; switch stuck open; wire makes loose contact; loss of electricity supply; faulty continuity in the electrical continuity of the wire	No light
Lamp	Directing the light flow	Does not allow for the correct orientation of the light flow	Arm can no longer be adjusted; lamp base is unstable; shade is degraded; user does not use it correctly	Does not allow for the correct orientation of the light flow
Lamp	Light bulb	Light bulb does not light up anymore	Primary cause	Does not produce light
Lamp	Arm	Lamp arm cannot be oriented anymore	Primary cause	Does not allow the correct orientation of the light flow
Lamp	Base	Base is unstable	Primary cause	Does not allow the correct orientation of the light flow
Lamp	Shade	Degraded lamp shade	Primary cause	Does not allow the correct orientation of the light flow
Lamp	Switch	Switch stuck open	Primary cause	Does not produce light
Lamp	Plug	Loose plug contact	Primary cause	Does not produce light
Lamp	Cord	Stripped wire	Unplugging by pulling the cord	Electric shock
Lamp	Cord	Loss of electricity continuity of the wire	Primary cause	Does not produce light
User	Plugging/Unplugging the lamp	Unplugging by pulling the wire	Not following the instructions	Stripped wire; wire with loose contact
User	Changing the light bulb	Changing the light bulb without following the instructions	Lack of training	Contact with a burning light bulb
User		Not following the instructions	Lack of training	Unplugging by pulling the cord
Environment	Electric protection	Fault in electric protection	Breakdown of the differential circuit-breaker	Electric shock
Environment	Circuit-breaker	Differential circuit breaker	Faulty relay; Faulty current detector	Faulty electric protection
Environment	Circuit-breaker	Faulty relay	Primary cause	Breakdown of the differential circuit-breaker
Environment	Circuit-breaker	Faulty current sensor	Primary cause	Breakdown of the differential circuit-breaker
Environment	Electricity supply	Loss of electricity supply	External cause	No light
Environment	Wire	Faulty continuity in the wire of the installation	Primary cause	No light

Figure 9.15. *FMEA example for the lamp*

9.6.2. *Chemical process*

The manufacturing process described in Appendix 7 uses, in its improved version, a group of two pumps. They are driven by a relay shown in Figure A7.2.

The FMEA analysis of this pumping function is shown in Figure 9.16. The failure modes of each of these elements are reviewed and, even for a small part of the installation, the FMEA analysis leads to relatively long tables. This information will be used to build the failure tree of the global system because the causes identified are primary causes for which it is possible to obtain reliability data. The failure tree obtained is shown in Figure 12.13.

System	Function or entity	Failure mode	Causes	Effects
Pumping station	Feeling the pumps	No electrical supply to pumps	Current wire disconnected [strong current wire]; relay stuck open [Relay]; relay is not controlled when closed [controlling the relay when closed]; Faulty open fuse [Fuse]; Breakdown [380 V supply]	Pumping station breakdown [pumping the fluid]
Pumping station	Controlling the relay to close	Relay not controlled in closed mode	Switch button stuck open [switch button]; Operator's error (does not press) [Operator]; breakdown [12 V supply]; Weak current wire disconnected [weak current circuit wire]	No electricity supply [feeding the pumps]
Pumping station	Pumping the fluid	Mechanical breakdown of two pumps	Mechanical breakdown 33041 [P33041]; Mechanical breakdown 33040 [P33040]	Pumping station breakdown [pumping the fluid]
Pumping station	Pumping the fluid	Pumping station breakdown	Mechanical breakdown of two pumps [pumping the fluid]; no electricity supply [feeding the pumps]	Problem with the cooling procedure
Pumping station	Switch button	Switch button stuck on	Primary cause	Relay not controlled when closed [controlling the relay when closed]
Pumping station	12 V supply	Faulty 12 V supply	External cause	Relay not controlled when closed [controlling the relay when closed]
Pumping station	Relay	Relay stuck open	Primary cause	No electricity supply for the pumps [feeding the pumps]
Pumping station	Weak current circuit wire	Weak current wires disconnected	Primary cause	Relay not controlled when closed [controlling the relay when closed]
Pumping station	380 V supply	Faulty 380 V supply	External cause	No electricity supply for the pumps [feeding the pumps]
Pumping station	Fuse	Faulty fuse open	External cause Reactivation error	No electricity supply for the pumps [feeding the pumps]
Pumping station	Strong current wire	Strong current wires disconnected	Primary cause	No electricity supply for the pumps [feeding the pumps]
Pumping station	P33041	Mechanical breakdown 33041	Primary cause	Mechanical breakdown of two pumps [pumping the fluid]
Pumping station	P33040	Mechanical breakdown 33040	Primary cause	Mechanical breakdown of two pumps [pumping the fluid]
Pumping station	Operator	Operator's error (does not press)	Human error (omission)	Relay not controlled when closed [controlling the relay when closed]

Figure 9.16. *Sample of the FMEA of the chemical process for the function "pumping"*

Chapter 10

Deviation Analysis Using the HAZOP Method

10.1. Introduction

The hazard and operability analysis (HAZOP) study method was developed by ICI for the chemical industry in 1963 [KLE 06]. The first publications describing this method date back to the 1970s [LAW 74]. The Union of Chemical Industries (UCI) published a French version of this method in 1980 in its safety specifications no. 2 entitled "Safety study on flowsheet". The principle of this method consists of systematically analyzing all the deviations of the operating parameters of the different elements or of the stages of the operating mode and in analyzing those that can potentially lead to a dangerous event. It is, in particular, well suited to systems that put into play material and energy flows. These flows are characterized by parameters such as flow speed, temperature, pressure, level and concentration; we examine the deviations of these parameters. Regarding its nature, this method is based on the analysis of the process flowsheets and the piping and instrumentation diagram (PID).

10.2. Implementation of the HAZOP method

The aim of the HAZOP method is to identify the risks and to study their prevention/protection by relying on the systematic analysis of all the possible deviations of the different parameters. The main stages of the method are as follows:

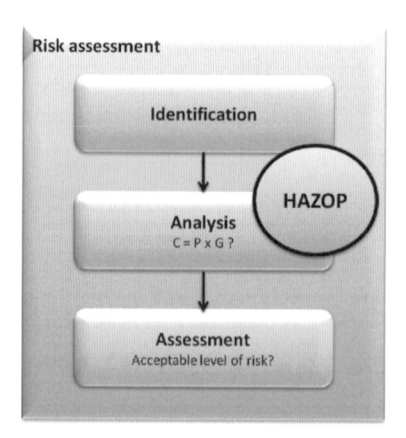

Figure 10.1. *The HAZOP method in the risk assessment process*

1) Analysis preparation: defining the context, obtaining information and drawing observations.

2) Description and installation modeling:

    i) study node definition;

    ii) characterization via variables.

3) Applying the HAZOP procedure:

    i) selection of the deviations that have damaging consequences;

    ii) effect analysis;

    iii) cause analysis;

    iv) *probability assessment (optional)*;

    v) *severity assessment (optional)*.

4) Definition of the actions that must be implemented, writing up the analysis report.

10.2.1. *Preparing the study*

The documents used for carrying out a HAZOP study often contain:

– process instrumentation diagrams;

– flow sheets;

– implementation diagrams.

We also use:

– operating handbooks;

– the sequential function chart of the process control system;

– description of safety instrumented systems;

– software description;

– sometimes, we will also use the handbooks describing different equipment.

This study is often carried out by a team made up initially of a team leader, a scribe and the people who have a good knowledge of the process and its design, for example the person responsible for the project, a process engineer, the (future) operation engineer, the instrumentation engineer, a chemical engineer and a maintenance engineer.

10.2.2. *Analysis of the study nodes*

The analysis of the study nodes is done by following the next stages (Figure 10.2):

1) In the first stage, choosing a study node. This generally encompasses a piece of equipment and its connections, and was identified during the modeling stage.

2) Choosing a operating parameter.

3) Selecting a keyword and generating a deviation.

4) Examining the consequences of this deviation. If they are not damaging the system, going to point 9.

5) Identifying the potential causes of this deviation.

6) Examining the means for detecting this deviation, as well as those designed to prevent the occurrence of similar deviations or limiting the effects of the considered deviation.

7) *Assessing the severity and the likelihood of this deviation (optional).*

204   Risk Analysis

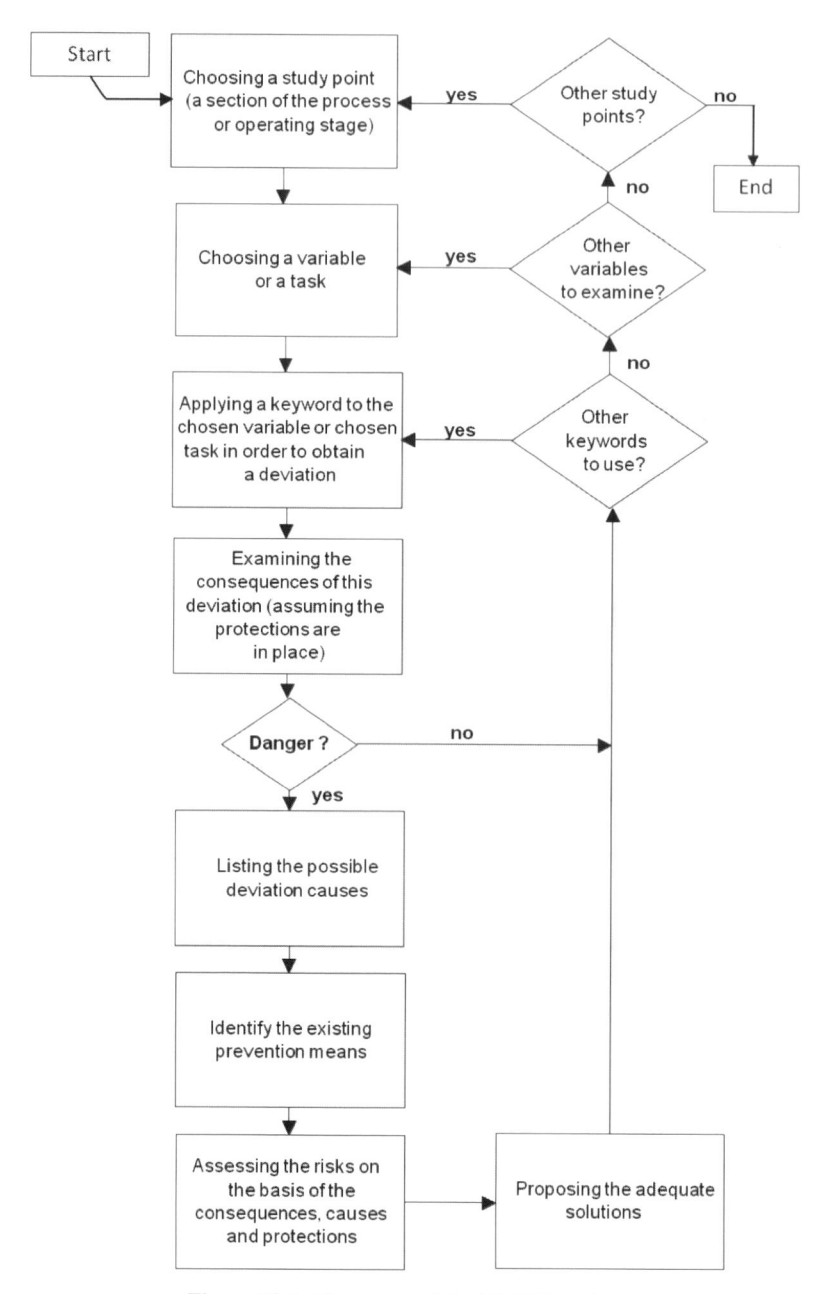

Figure 10.2. *The stages of the HAZOP method*

8) Proposing, should the situation require it, recommendations for improvement.

9) Retaining a new keyword for the same parameter and restarting the analysis from point 3.

10) When all of the keywords have been considered, retaining a new parameter and restarting the analysis from point 2.

The points of study that are characterized by variables are mainly equipment parts, such as tanks, reactors, distillation unit, mixing tanks, pumping units, compressors, etc. These points of study may also be operating steps.

KEYWORDS	APPLICATION	SIGNIFICATION
MORE	Process parameters Measurable quantities Actions	Value too high
HIGH		Value above the threshold
LESS		Value too low
LOW		Value below the threshold
NONE		Null value
INVERSE		Opposite sense, Inverse operation carried out
NO	Actions	Forgotten, omitted
ALSO		Another operation carried out with the normal operation
PARTLY		Normal operation carried out incompletely
OTHER THAN		Another operation carried out
NOT AT THE RIGHT TIME		Before, After
LEAKAGE	Containers, Piping	Leakage on the element
FAILURE		Failure of the element

Figure 10.3. *Keywords*

A list of typical keywords is given in Table 10.3. These keywords are combined with the variables characterizing the study nodes in order to provide what we call *deviations*. These deviations are, for example:

– If the viewpoint is a pipe, we will be able to characterize it by the variables *flow, speed* and *pressure*. By applying the keywords, we will obtain, for example:

NO + FLOW = NO FLOW or TOO MUCH + PRESSURE = TOO MUCH PRESSURE

– If the study nodes are operating steps, we will be able to obtain:

MORE + A PHASE = TWO PHASES or ANOTHER + EXPLOITATION = MAINTENANCE

Certain combinations do not have any sense and are therefore not used. When the deviation is being generated, the work group studies the possible consequences of this deviation. If no dangerous consequence is found, we may move on to the next deviation. We will note that, in this case, the deviation is not mentioned in the results table.

The parameters used generally concern the pieces of equipment. We find values such as (Figure 10.4):

– the process parameters: P, T, L, F, [C], speeds, etc.;

– measurable quantities: charges, durations, etc.

List of standard parameters	
- Temperature	- Frequency
- Flow	- Viscosity
- Pressure	- Tension
- Level	- Information
- Time	- Agitation
- Composition	- Addition
- pH	- Separation
- Speed	- Reaction
- Residence time	- Concentrations
- Impurities	- Physical properties
- Containment	- Utilities

Figure 10.4. *List of standard parameters*

The parameters can also represent actions of the operating system such as: filling in, warming up, cooling, emptying, etc. Specific keywords are associated with these actions.

10.2.3. *Causes and consequences of the deviation*

The consequences of the deviations are dangerous phenomena that are often encountered in the chemical industry: fire, explosion and toxic dispersion. The causes of the deviations are often due to failures in the equipment or human error.

	More High	Less Low	Not at all None	Also	Partly	Inverse	Other than	Before	After
Pressure	x	x	x						
Temperature	x	x							
Level	x	x	x						
Flow	x	x	x			x			
Frequency	x	x							
Viscosity	x	x							
Tension	x	x	x						
pH	x	x							
Time	x	x							
Agitation	x	x	x		x	x	x	x	x
Speed	x	x	x			x			
Residence time	x	x							
Concentrations	x	x	x	x			x		
Impurities	x	x	x	x			x		
Physical properties	x	x							
Containment			x						
Utilities			x						
Information			x		x		x	x	x
Addition	x	x	x		x		x	x	x
Separation	x	x	x		x				
Reaction	x	x	x		x		x	x	x

Figure 10.5. *List of possible combinations*

Here are several examples of causes for a deviation of a *high level* type:

– opening/closing the regulation valve: sensor failure, computer failure, actuator failure, valve failure, set point error, etc.;

– exceeding normal supplies: excessive valve opening;

– insufficient outputs;

– valve closing;

– obstruction, clogging, foreign material, etc.;

– undesirable supply via another circuit: examining all of the lines on the device;

– liquid input: water, vapor, oil, etc.;

– temperature or pressure variation;

– abnormally low density;

– foaming, condensation, etc.;

– operator error.

For *low level-type deviations*, we have:

– opening/closing the regulation valve: sensor failure, computer failure, actuator failure, valve failure, set point error, etc.

– exceeding normal outputs: excessive valve opening;

– insufficient supply;

– valve closing;

– obstruction, clogging and foreign material;

– undesirable outputs by another circuit: examining all the device lines;

– outputs via utility circuits: water, vapor, oil, etc.;

– leak toward the outside;

– pressure or temperature variation;

– abnormally high density;

– vaporization and evaporation;

– operator error.

10.2.4. *Result tables*

Recording the results is a significant part of the HAZOP reviews. In general, the deviations that do not have any consequences on safety are not registered. The results are generally provided in a tabular format (Figure 10.6). The assessment of the likelihood and the severity are not always indicated.

10.3. Limits and connections with other methods

The HAZOP method is limited to the systems whose state we can characterize by a set of physical–chemical variables. It cannot be applied to a mechanical system, for example. Moreover, it does not explicitly consider all of the failures of a system, but only those whose effects result in a deviation of the operating state. This is an advantage because this method needs fewer resources and is less tedious, but it can also be limiting since certain aspects may not be taken into consideration.

System	Entity Function	Deviation	Causes	Consequences	S	P	Detection means	Existing prevention actions	Notes
(1)	(2)	(3)	(4)	(5)	(6)	(7)	(8)	(9)	(10)

Figure 10.6. *HAZOP table*

EXAMPLE 10.1.– If we are interested in flow deviation (NOT ENOUGH FLOW), we treat all of the failures that trigger this deviation globally, such as a supply failure, blocked valves, clogged pipe, etc.

10.4. Model-based analysis

In a way that is similar to the preliminary hazard analysis (PHA) or the failure mode, effects and criticality analysis (FMECA), the HAZOP method may be guided by a model. The consequences of a deviation can be expressed by following the typology described in section 7.9, namely:

– a dangerous event linked to a system, or to another function or entity, that is identified via the PHA method;

– a failure mode linked to a system, or to another function or entity, that is identified using the FMECA method;

– another deviation, such as a deviation in temperature, that triggers a pressure deviation;

– an event that characterizes a degradation.

The same applies for the causes. The links between the different methods, therefore, appear quite clearly, and the scenarios can be obtained directly.

10.5. Application example

To illustrate the method, we will consider the example of the exothermic reactor described in Appendix 7. The HAZOP analysis is presented in Table 10.7. The study nodes considered are:

- the reactor;
- supply line A;
- line B;
- the cooling system;
- the operator.

The analysis is more concise than the analysis carried out with the FMECA method. Not all the failures and the effects can be expressed as deviations and we therefore find elements of the PHA analysis for the effects, and elements of the FMECA analysis for the causes.

HAZOP				
System	Function or Entity	Deviation	Causes	Effects
Reacting system	R33030 Reactor	Too high temperature	Insufficient cooling Not enough agitation Other than Composition A Other than Composition B Overload	Too much pressure
Reacting system	R33030 Reactor	Too much pressure	Too high temperature Blocked valve	Explosion and population intoxication
Reacting system	R33030 Reactor	LEAK	Corrosion	Staff intoxication
Reacting system	Stirrer	Not enough agitation	Motor breakdown Electric supply loss Command error	Too high temperature
Reacting system	Line A	Other than Composition A	Supplier error	Too high temperature
Reacting system	Line B	Other than Composition B	Supplier error	Too high temperature
Cooling system	Cooling line	Not enough flow	Industrial water supply loss	Insufficient cooling
Cooling system	Cooling line	Too high Temperature	Regulation problem	Insufficient cooling
Operator	Load	Overload	Human Error	Too high temperature

Figure 10.7. *Extract of the HAZOP analysis for the chemical process*

Chapter 11

The Systemic and Organized Risk Analysis Method

11.1. Introduction

The SORAM method [PER 07] is a systemic and organized risk analysis method that was developed at the CEA in the 1980s. It is based on the system dysfunctions analysis methodology (SDAM) model, developed in the 1980s [LES 02] and shown in Figure 11.1. It describes what is called a "danger process" in the SDAM-SORAM terminology. The hazard source generates a danger flow that produces damage on a target system. The set source-target is influenced by a danger field, which creates particular conditions that either limit or amplify the danger flow. The danger flow may be produced after the occurrence of initiating events, which can be either internal or external. The damage done upon the target are described by a final event.

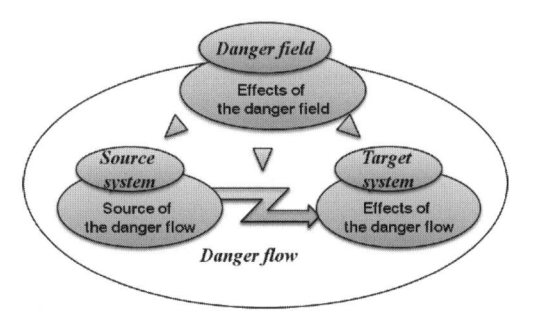

Figure 11.1. *SDAM model*

Figure 11.2 shows an event-based view of the SDAM model. The dangerous phenomena are the events corresponding to the danger flows. The causes for the occurrence of dangerous phenomena are initiating events that are either internal or external. The damages are represented by final events on the target. The target can itself become in turn a source of danger, the consequence of the flow being a dangerous phenomenon, source of another danger flow. This method brings about the connections between the danger sources of the different systems that compose an installation and is therefore well suited to the study of domino effects.

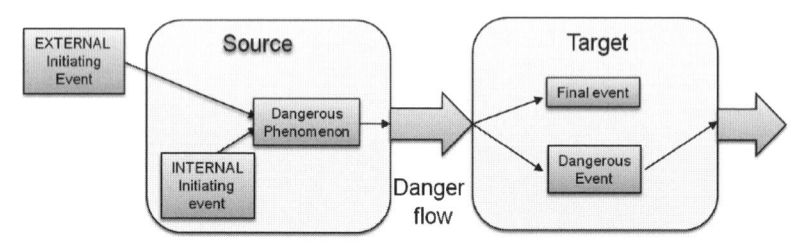

Figure 11.2. *SDAM model in event-like form*

The SORAM method is made up of two modules that can be used relatively independently from one another. Module A corresponds to a macroscopic risk analysis and is similar to a preliminary risk analysis. Module B corresponds to a more detailed analysis of the scenarios identified within module A. The SORAM method can be seen as a reformulation of the classical risk analysis methods based on a systemic modeling of the installation.

It also proposes analysis grids that help identify the risks. These are danger source lists that are classified into the following eight categories:

– Hazard source systems of a mechanical origin.

– Hazard source systems of a chemical origin.

– Hazard source systems of an electrical origin.

– Fire hazard source systems.

– Hazard source systems linked to radiation.

– Hazard source systems of a biological origin.

– Human-related hazard source systems.

– Hazard source systems related to the active environment.

Module A comprises the following stages:

1) Modeling of the installation: this stage consists of representing the installation via a set of interacting systems.

2) Identification of the danger sources: this is a stage where each system is analyzed using a grid in order to determine to which danger source category it belongs, and in order to establish the danger model of the system, thus called a "black box".

3) Scenario construction: this stage consists of combining danger models in order to generate chains of events leading up to damages and obtaining logical pre-trees.

4) Assessment of the severity of the scenarios: this stage consists of assessing the severity of the final event(s).

5) Negotiation of the objectives: this stage consists of defining a risk matrix and selecting the acceptable or unacceptable cells.

6) Qualitative definition of the safety barriers: this is a stage where a certain number of barriers are proposed in order to lower the risk level of the scenarios deemed as unacceptable.

Module B allows us to refine the study by producing more detailed scenarios and quantifying the respective risks. It follows the next stages:

1) Dysfunction identification is a stage that consists of identifying the failure modes and/or the potential deviations that can appear in the scenarios, using either an FMECA or an HAZOP method.

2) Building a fault tree on the basis of the modes identified in the previous stage, in order to numerically assess the probability of the scenario.

3) Negotiating precise objectives with the new value of the scenario probability that is calculated from primary causes.

4) Barrier quantification: this quantification is used in order to calculate the criticality of the scenarios with the barriers.

A certain number of these stages are not specific to the SORAM method, particularly stages A4, A5, A6, B1, B2, B3 and B4. Therefore, in what follows, we will mainly present the specific stages.

214   Risk Analysis

The progressive and permissive aspects of the SORAM method are undeniable advantages, but its implementation sometimes turns out to be difficult because the application means are not formalized [GAR 99] and because there are no writing and formalizing rules for the modeling of the accidents' scenarios in the shape of black boxes [FUM 01].

In what follows, we will present a formalized version of the method [FLA 01] which allows us, by using a computerized tool (Appendix H), to generate automatically the preliminary hazard analysis (PHA), the FMECA analysis and the fault trees associated with the SORAM analysis.

11.2. Implementation of part A

11.2.1. *Modeling of the installation*

In the SORAM method, the installation is represented as systems, sometimes referred to as subsystems (Figure 11.3). Most of the time, we add a particular system called an "environment" that allows us to represent what surrounds the installation. In the case of normal functioning, a certain number of interactions take place between these systems. They are classified into three categories:

– material flows;

– energy flows;

– information flows.

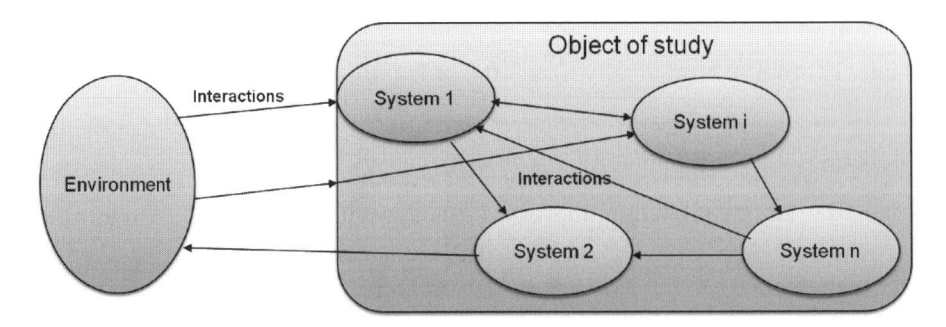

Figure 11.3. *Flows between the systems*

REMARK 11.1.– To carry out the risk analysis, it is only required to decompose the installation into subsystems. It is not necessary to carry out a complete inventory of the flows between the systems, which is mainly an aid in understanding how the installation works.

EXAMPLE 11.1.– We will present the implementation of the method on the simple pedagogical example that we have already used (section 8.5.1) in the previous chapters: a desk lamp and its environment. To represent the lamp in its usage context, there follows a list of systems that we will consider:

– S1: the lamp;

– S2: the user;

– S3: the environment (the rest).

In the normal operating mode, a certain number of flows can be identified:

– Lamp light flow toward the environment and the user.

– Electricity flow toward the lamp.

– Command information flow from the user toward the lamp.

As specified above, these flows are not used for the analysis, but their search serves for understanding the functioning of the whole.

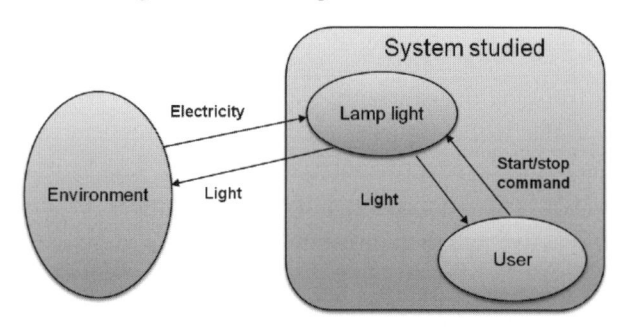

Figure 11.4. *Modeling of the lamp*

11.2.2. *Identification of the hazard sources*

The objective of this stage is to identify, for each of the previously identified systems, the hazard sources that may generate damage, the targets and the situations in which these damages take place. This phase is close to a PHA. It is based on a predefined list of hazard sources. We analyze each system by using this list and, if the system belongs to a source type, we identify the danger flows it can generate, and then the conditions in which this danger flow can be generated. The danger flows correspond to dangerous phenomena that can bring about damages, either on the system that is being analyzed or on other systems. The result of the analysis is given in a table called Table A (Figure 11.5).

SORAM Table A – System :						
Source type	Lifecycle phase	External initiating event	Internal initiating event	Main event	Final event	Target

Figure 11.5. *Table A for system analysis*

This table must be filled in as follows:

1) Selecting a system.

2) Starting from the list of hazard sources (Figure 11.6), we must analyze the system in order to determine if it belongs to this source category. For example, is the system a hazard source of type A11, i.e. "is it a device under gas pressure?"? If it is, we must identify in which phase of the lifecycle this hazard source is active and then move on to stage 3.

3) We must fill in the column "Source category" (in this case, A11), and then the associated danger flow, for example "Explosion".

4) Looking for possible causes: these causes will be represented by an initiating event that can be either internal or external to the system.

5) Defining the target and the effect upon the target, herein called final event.

A: Mechanical hazard sources
A.A1: Pressurized devices
    A11-Gas: device subject to gas pressure
    A12-Vapor: device subject to vapor pressure
    A13-Liquid: device subject to hydraulic pressure
A2: Elements subject to mechanical stress
A3: Moving elements
A4: Elements requiring handling
    A41: Elements requiring manual handling
    A42: Elements requiring mechanical handling
A5: Sources of physical explosions
A6: Fall from height hazards
A7: Sources of slip, and fall on the same level
A8: Sources of injuries by inappropriate contact
A9: Sources of noises and vibrations
B: Chemical hazard sources
  B1: Sources of uncontrolled chemical rections
  B2: Sources of explosion
  B3: Sources of toxicity and corrosiveness
  B4: Sources of pollution and odors
  B5: Sources of lack of oxygen
C: Electrical hazard sources
  C1: DC or AC electrical current
  C2: Static electricity
  C3: Capacitors
  C4: High frequencies
D: Fire risk sources
E: Thermal and radiation causes of danger
  E1: Ionizing radiation
  E2: Heat sources
    E21: Heat conduction
    E22: UV / IR / visible radiation
  E3: Laser radiation
  E4: Microwave radiation
  E5: Magnetic field
F: Biological hazard sources
  F1: Viruses / Bacteria
  F2: Toxins
G: Human as hazard source
  G1: Normal situation
  G2: Malicious actions
H: Hazard sources relating to the active environment

Figure 11.6. *SORAM hazard source grid*

218   Risk Analysis

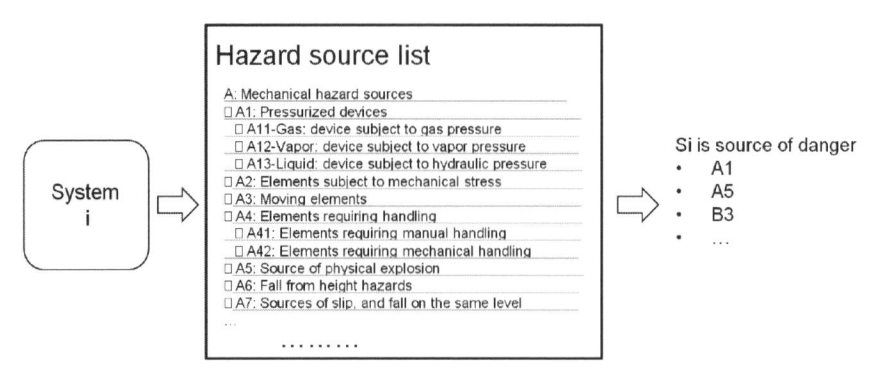

Figure 11.7. *Analysis process*

The common phases of the lifecycle are:

– running (EX);

– maintenance (MT);

– start-up (UP);

– stop (SW).

Let us note that this table highlights portions of scenarios that take place either internally or externally to a system and that describe the causal relations as follows:

causes → hazard flow (or main event) → effects

for each risk identified. In this table, the barriers are not indicated. A review of the barriers is carried out in the second stage.

Table A allows us to build the black box model of the system. The main outputs of this box are danger flows generated by the system, the inputs being the flows received from the other systems (Figure 11.8). The other events, which are generated by the system, but which are not danger flows, can also be represented. It is important that we build the table sparingly. It is useless to select all the hazard source categories for a system, if they are not relevant. Because the analysis is iterative, and the black boxes can be completed during the building of the scenarios, it is always possible to complete the model.

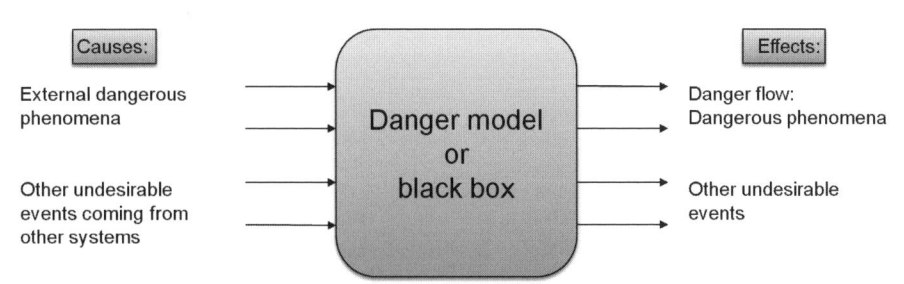

The Systemic and Organized Risk Analysis Method 219

<u>Dangerous phenomena (or Danger flow)</u> : is a physical flow (material, energy) or informational which corresponds to an undesirable exchange responsible for a direct degradation of the target by its action.

Figure 11.8. *Danger model or black box*

EXAMPLE 11.2.– Let us take the example of the desk lamp again. Let us consider that the S01 system represents the lamp. This system belongs to four hazard source categories:

– A8: direct injury source, generating a lesion via direct contact in the case of a wound caused by jamming.

– C1: electricity source, which can lead to electric shock.

– D: source of fire.

– E21: thermal source (via conduction), leading to burns.

For each of these sources, an analysis is carried out in order to identify the danger flows and their initiating events. The list of the identified danger flows is as follows:

– Shock, causing injuries in case of untimely collision.

– Electric shock, which can be caused by a failure of the electrical protection.

– Burns, which can take place in the maintenance phase when the light bulb is being changed.

– Overheating leading to a fire caused by poor ventilation.

This work must be carried out for each system. The result is given in Figure 11.9 for the "lamp" system and in Figure 11.10 for the "user" system.

SORAM Table A – S01: Lamp

Source type	Lifecycle phase	External initiating event	Internal initiating event	Main event	Final event	Target
A8:Source of direct injuries	EX	Unexpected collision		Shock	Wounds	User
E21:Thermal conduction	MT	Lightbulb changing Without the necessary precautions		Burns	Wounds induced by burning	User
C1:Electricity based on direct or alternating current	EX	Failure of electrical protection	Stripped wire	Electric shock	Death	User
D:Fire hazard source	EX		Poor ventilation	Heating	Fire	Environnement

Figure 11.9. *Table A for the "lamp" system*

SORAM Table A - User

Source type	Lifecycle phase	External initiating event	Internal initiating event	Main event	Final event	Target
G: Human as hazard source	EX		Wrong gesture	Unexpected collision	Shock	Lamp
G: Human as hazard source	EX		Does not follow instructions	Turning off by unplugging	Stripped wire	Lamp
G: Human as hazard source	MT		Lack of training	Lightbulb change without necessary precautions		

Figure 11.10. *Table A for the "user" system*

11.2.3. *Building the scenarios*

Building the scenarios consists of combining the black box models when possible, that is when the output of a system corresponds to the input of another system (Figure 11.11). This is the case, for example, if a system is a source of danger that generates shocks, i.e. has a "shock"-type danger flow for an output, and another system is sensitive to this flow, i.e. its input is an external initiating event of "shock"-type danger flow. These combinations allow us to develop logical pre-trees, which can be seen as bow-tie diagrams, without the AND and OR gates. When the final event and the initiating events belong to the system that generates the danger flow, we may call it a short scenario, whereas in the other cases, we may call it a long scenario.

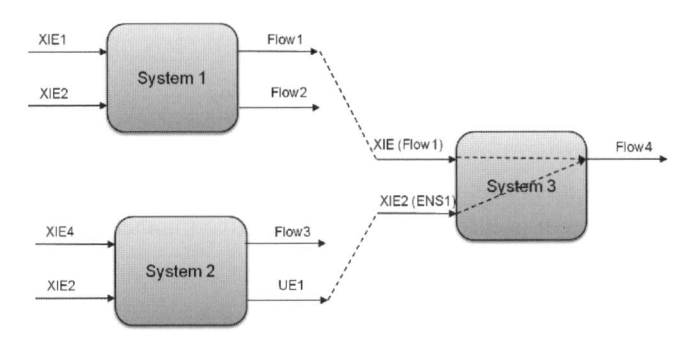

Figure 11.11. *Building a scenario*

In practice, during this stage, the black box models are modified and adapted. Certain causal relations appear more clearly when building the logical tree (Figure 11.12). Therefore, it is easier to work from this tree perspective and to update black box models depending on the tree we have built.

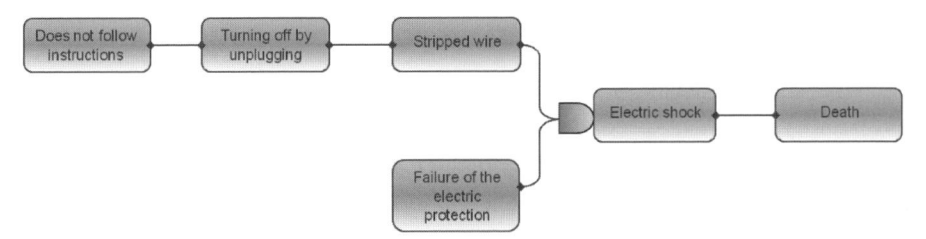

Figure 11.12. *Scenario 1 corresponding to electric shock*

After this construction, we have the possibility of bringing together in the same graph all the events and pathways leading up to a given undesirable event, such as an accident. The logical operators, OR and AND, will have to be specified at a later stage in order to convert this graph into a fault tree or into a bow-tie diagram.

EXAMPLE 11.3.– The main event "electric shock", corresponding to the third row of Table A of the "lamp" system, leads to a "death". The causes of this event are given by the internal and external initiating events. The event "stripped wire" can be found in the column "Final event" of Table A of the "user" system. The cause, or the main corresponding event, is a "turning off by unplugging". This event is itself caused by the event "does not follow the instructions". This analysis allows us to build the scenario[1] given in Figure 11.12.

11.2.4. *Assessment of the severity of the scenarios*

This stage consists of assessing the probability of the undesirable event and the severity of its consequences, for all the scenarios built in the previous stage. This assessment is carried out in a qualitative manner in stage A and in a quantitative manner in stage B.

In stage A, we choose the event that represents the damages, generally the final event, and we determine its severity in a qualitative manner. To do this, a severity scale is designed in the traditional way, as mentioned in section 6.3.

EXAMPLE 11.4.– Let us take the example of the desk lamp again. The scenario leading up to electric shock has a very high severity, potentially causing death. This level of damage will be placed on the chosen scale at level S1, following the process mentioned in Chapter 6.

11.2.5. *Negotiation of the objectives*

This stage consists of defining a qualitative scale of different levels of probability, in order to complete the severity scale. Then, starting from the risk matrix, we can negotiate the criticality levels that are deemed acceptable and those that are deemed unacceptable. An example is given in Figure 11.13. The levels defined are:

– A1: unacceptable;

– A2 : acceptable, but should be improved if possible;

– A3: tolerable.

This phase consists of defining a risk matrix, similar to the other methods. In the case of an implementation of risk management in the ISO31000

---

[1] In the SORAM method, the term "scenario" has a different meaning from the usual one. In fact, it indicates the logical tree that represents all the possible scenarios causing a final event.

standard, the matrix and the decision criteria will have been defined in the task of context establishment.

Severity						
S1	A2	A2	A1	A1	A1
S2	A2	A2	A1	A1	A1
S3	A3	A2	A2	A1	A1
S4	A3	A2	A2	A2	A1
S5	A3	A3	A3	A2	A2
	E	D	C	B	A

Figure 11.13. *Probability-severity matrix*

EXAMPLE 11.5.– Scenario 1 in our example can either be considered probable or as having a probability level B. The table gives us an A1 acceptability value for the couple (B, S1). This scenario is, therefore, unacceptable.

11.2.6. *Proposing the barriers*

Having analyzed the scenario, this stage consists of defining the barriers for preventing or limiting the occurrence of the final event of the scenario.

In the SORAM method, the barriers are classified into the following two categories:

– Technical barriers (TB): these are technological elements or solutions integrated in the installation and capable of reducing the probability or the severity of the scenario. These barriers can be static, meaning they can actually stop the occurrence of the main event, such as for example a protection hood or a screen, or they can be dynamic, i.e. act in reaction to the undesirable event, such as for example a safety valve, an automatic safety system or a circuit breaker.

– Human barriers (HB): these barriers rely on human behavior, such as following instructions, implementing the correct procedure, use of personal protective equipment (PPE) or a human intervention faced with an undesirable event, such as the implementation of an emergency procedure. They are generally considered less efficient than the technological barriers.

224   Risk Analysis

The set of barriers suggested is sometimes presented in a table called Table B (Figure 11.14). The scenarios are re-evaluated using the new barriers, with regard to both severity and probability. We can then verify that the criticality value reaches an acceptable level.

SORAM Table B					
Scenario No	Lifecycle phase	After event	Before event	Barrier	Type

Figure 11.14. *Barrier table*

EXAMPLE 11.6.– For our example, by limiting ourselves to scenario 1, we propose in the table, presented in Figure 11.15, a series of measures for limiting the occurrence of an "electric shock" event, both regarding the failure of the electrical protection and the disconnection and pulling of the wire. These measures allow us to reduce the probability of electric shock. A reassessment of this probability using the proposed measures allows us to estimate it to the value "Highly improbable" or D. The severity does not change. The criticality moves to level A2.

SORAM Table B					
Scenario No	Lifecycle phase	After event	Before event	Barrier	Type
1	EX	Not following the instructions	Switching off by pulling the plug	Wire reinforcement	BT
1	EX		Not following the instructions	Training	BU
1	MT		Circuit-breaker breakdown	Periodic verification of the circuit-breaker	BT

Figure 11.15. *Barrier table for the example of the lamp*

11.3. Implementing part B

The objective of part B is to resume the analysis by carrying out a quantified assessment of the criticality of the scenarios:

– completing them with a more detailed analysis of the events leading up to the main undesirable event and specifying if the logical connectors are of OR or AND type, which transforms the logical tree in a fault tree;

– quantifying the probabilities of the basic events of this tree and using the appropriate rules for calculating the probabilities.

11.3.1. *Identifying the possible dysfunction*

A certain number of initiating events that were introduced in the analysis in stage A must be explained so that it becomes possible to assess its probability. To do this, an FMECA or an HAZOP analysis can be carried out in order to analyze the causes of the failures or the deviations leading up to this initiating event. These failure modes or deviations will, therefore, allow us to build the failure trees leading up to initiating events. In the simple cases, they will replace the initiating events altogether.

EXAMPLE 11.7.– In our example, the FMECA analysis is based on the structural-functioning model given in Figure 7.13. Let us consider, for example, the functions that are connected to scenario 1:

– S2.F3: connecting/disconnecting the lamp, a function of the user system;

– S3.F1: ensuring electric protection, a function of the environment system.

Among these functions, a failure of the function S2.F3 corresponds to "disconnecting by pulling the wire", which is not in accordance with the instructions for connecting/disconnecting the device. A failure in function S3.F1, ensuring electrical protection, corresponds to "failure of electrical protection". The FMECA analysis allows us to identify two causes, either a "relay fault" for the circuit breaker, or a "sensor fault". These two causes allow us to add new information to the logical pre-tree corresponding to scenario 1.

11.3.2. *Building the fault tree*

Starting from initiating events that have already been identified in Table A, completed by those identified via the dysfunction analysis, we can proceed to building a fault tree. It takes the structure of the pre-tree again and adds to it the failure modes identified in the previous stage. Moreover, the AND and OR connectors are specified.

EXAMPLE 11.8.– Figure 11.16 shows the tree obtained for scenario 1 in our example.

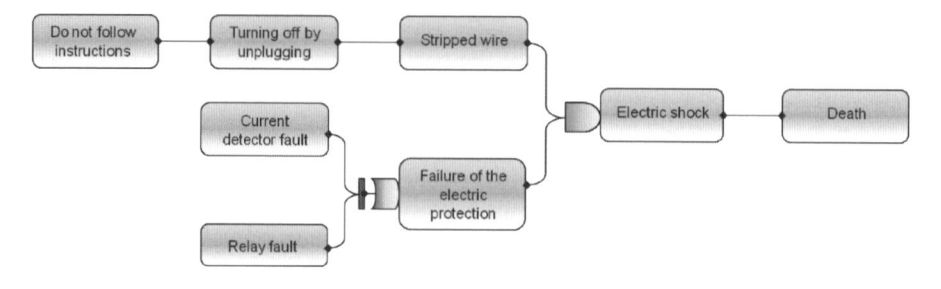

Figure 11.16. *Scenario 1 corresponding to the electric shock after FMECA*

11.3.3. *Negotiation of quantified objectives*

The risk matrix defined previously must be redefined for a probability and a severity represented by quantitative values. The simplest solution consists of defining a table of correspondence between these types of value. We can for example use the values proposed in section 5.6:

A: 1 to 10^{-2}, B: 10^{-2} to 10^{-3}, C: 10^{-3} to 10^{-4}, D: 10^{-4} to 10^{-5}, E: less than 10^{-5}

By using this equivalence, it is then possible to define a risk matrix using quantitative values.

11.3.4. *Barrier quantification*

The efficiency level of the barriers must also be quantified. For prevention barriers, which act on the probability, we must define the probability of failure on demand via confidence levels (CL), using the relation $PFD = 10^{-NC}$. This aspect is developed in Chapter 15. Usually, the integer values from 1 to 4 are used. Regarding the protection barriers acting upon the severity, we define a similar parameter called L.

The probability of an event with a barrier becomes: $P = P_0 10^{-NC}$ and its severity becomes $G = G_0 10^{-L}$.

EXAMPLE 11.9.– For our example, limiting ourselves to scenario 1, the measures proposed in part A are quantified (Figure 11.17). For each of them, we can then evaluate the probability and severity of the scenario in a quantitative manner (Figure 11.18).

SORAM Table B

Scenario No	Lifecycle phase	After event	Before event	Barrier	Type	NC	L
1	EX	Not following the instructions	Switching off by pulling the plug	Wire reinforcement	BT	0.01	
1	EX		Not following the instructions	Training	BU	0.1	
1	MT		Circuit-breaker breakdown	Periodic verification of the circuit-breaker	BT	0.1	

Figure 11.17. *Table of the quantified barriers for the example of the lamp*

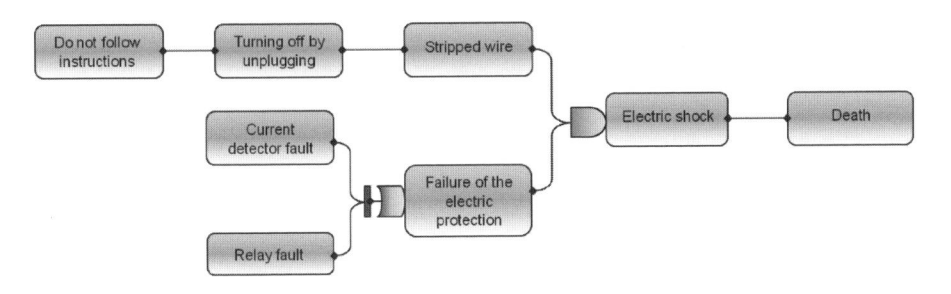

Figure 11.18. *Failure tree with the barriers included*

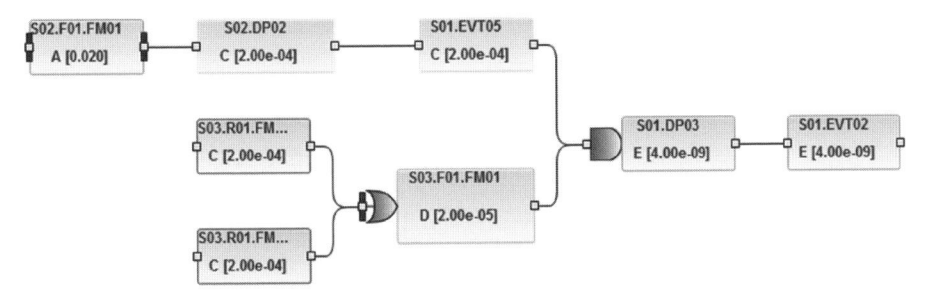

Figure 11.19. *Failure tree with the probabilities included*

11.4. Conclusion

The SORAM method is a complete risk analysis procedure. Some of its stages are based on other methods, and it can be seen as a way of organizing these methods in a consistent manner. It is an interesting method, although mostly it is limited to a partial implementation, namely to its Part A.

Chapter 12

Fault Tree Analysis

12.1. Introduction

A fault tree is a model that enables us to highlight the logical combinations of the faults that could lead to the event we are mainly concerned with, hereafter called "the top event". Its graphical representation takes the form of a tree (Figure 12.1). In general, this event brings about significant damage that has been previously identified via other methods (preliminary hazard analysis (PHA), failure mode and effects analysis (FMEA), hazard and operability study (HAZOP), etc.), and we use the fault tree approach to analyze in more detail the basic event combinations that could lead to this event, to assess the probability that such an event might occur and to reflect on the positioning of the safety barriers.

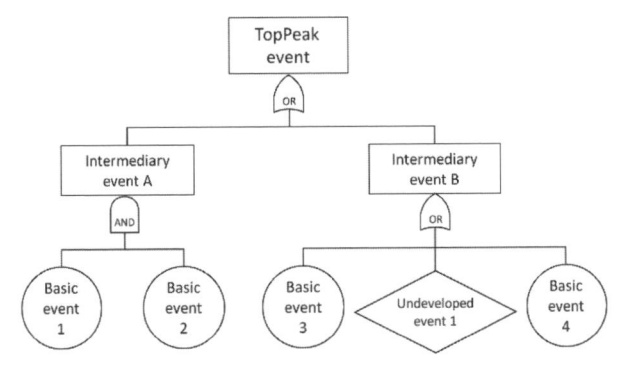

Figure 12.1. *Example of a fault tree*

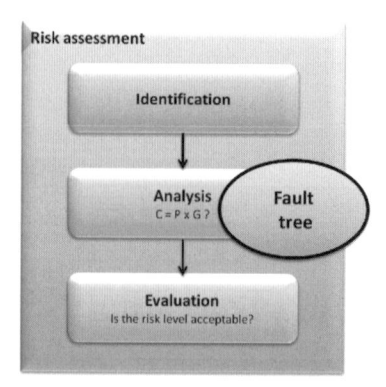

Figure 12.2. *The fault tree in the risk assessment process*

The method was carried out by Bell Telephone laboratories in 1961, and used for the risk analysis of firing systems in intercontinental missiles. It was further improved by Boeing and used for the safety analysis of nuclear reactors [NUR 75]. It has been used in several other fields since then.

12.2. Method description

The analysis via fault trees is a deductive method: starting from a given event, we are concerned with identifying all the combinations that could lead to that event. This analysis is carried out by going from cause to cause, until we reach events that are deemed elementary, that is events that will not be explained any further. Usually, these elementary events correspond to:

– events that represent simple facts, such as equipment faults, for which an estimation of the probability is indeed possible;

– events representing facts that normally take place during the system operation;

– events that cannot be deemed elementary, but that we choose not to analyze in more detail because of a lack of interest or because of insufficient information.

The causes are connected by a logical operator to the event that is being analyzed. In the majority of these cases, this operator is an OR logical operator or an AND operator, but there are also other possibilities. Once the tree is built, it offers a graphical overview that allows us to quickly grasp the conditions that the top event needs in order to take place. The fault tree allows us to qualitatively represent the occurrence conditions of an event. However, it can also be used to calculate the probability of the top event: the basic events are used to form combinations that could generate the top event, which in turn allows us to calculate its probability. For this calculation to be possible, the basic events must verify the following properties:

– they must be independent;

– their probability must be known.

12.3. Useful notions

12.3.1. *Definitions*

DEFINITION 12.1.– *The faults and the failures of a function or component can be classified into three categories, called primary-secondary-command (PSC) categories:*

– Primary faults/failures: these are internal to the entity or to the function. It is, for example, the case of the fault of a pump due to the splitting of the tree.

– Secondary faults/failures: these are faults or failures that are due to the external elements that are used by the entity or due to external conditions, such as, for example, the fault of a pump caused by a liquid that is either too viscous or too heavily charged with impurities, causing the pump to degrade.

– Faults and failures of the command: these are faults or failures that are also external, but due to a command error of the entity or the function. For example, for the pump, a start-up command that has not been sent.

The aim of this classification is to help study the causes of an event when we build the tree: we first seek to find out the primary causes, then the secondary ones, then the command ones if the entity is indeed being commanded by someone.

DEFINITION 12.2.– *The Immediate, Necessary and Sufficient (INS) causes are made up of a set of immediate causes, both necessary and sufficient, that could make the event happen:*

– A cause is called immediate if it directly precedes the event in the causality chain, having considered the chosen description level.

– *A cause is necessary if the event being analyzed cannot take place if the cause does not take place as well.*

– *A set of causes is said to be sufficient if the occurrence of the causes triggers the occurrence of the event, immediately or at a later date, within the normal activity. The event can take place upon the occurrence of a normal event, such as starting up an element. This normal event is generally not included in the fault tree.*

To determine the INS causes, it is useful to rely on a model to guide the analysis. For example, if we use a structuro-functional model, we know that the cause of the failure of a function is either a failure of an element connected as input or an undesirable event. The identification of the causes is then easier.

DEFINITION 12.3.– *A cut set is a set of basic events that, when they are simultaneously active, trigger the occurrence of the top event. A cut set is said to be minimal if it is impossible to retain one of its elements without the entire set being a cut set.*

12.3.2. *Graphical representation of events and connections*

The events are represented by symbols that are shown in Figure 12.3. We can distinguish several types of representation:

– basic events, which we consider to be elementary;

– the intermediary events that result from the logical combination of other events;

– the events that are not developed within the study;

– the events that are connected to normal activity.

The gates (Figure 12.4) allow us to represent the logical connections between the causes of an event.

REMARK 12.1.– A fault tree contains logical information, that is it describes the connections between an event and the causes it needs in order to take place. It does not describe the temporal aspects: the causes do not necessarily trigger the immediate occurrence of the event.

Fault Tree Analysis   233

Symbol	Name	Significance
 (circle)	Basic event	This event is an elementary event that does not need explaining.
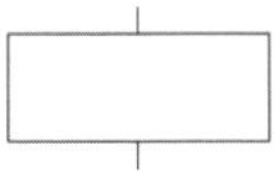 (rectangle)	Intermediary event	This event is generated by a connection.
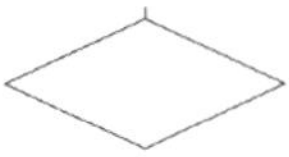 (diamond)	Undeveloped event	This event is not elementary, but it is not developed because it is beyond the scope of the study or because of a lack of information.
 (house)	External basic event	The event is supposed to take place during the normal functioning of the system.

Figure 12.3. *The different types of events*

Symbol	Name	Significance
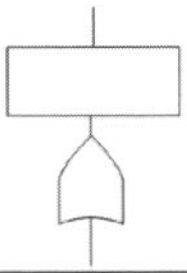 (OR gate symbol)	OR gate	The output event takes place if any of the inputs events take place.
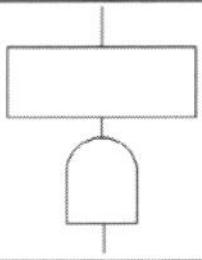 (AND gate symbol)	AND gate	The output event takes place if all the input events take place.
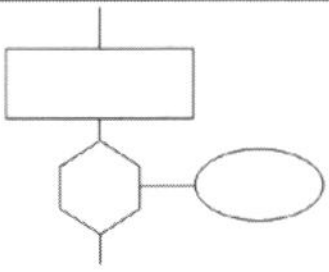 (Inhibition gate symbol)	Inhibition gate	The output event takes place when the input event takes place if the condition (represented in the oval) is true.
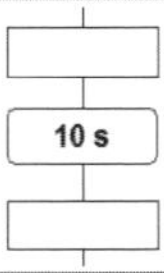 (Delay gate symbol, 10 s)	Delay gate	The output event takes place with the delay of the duration indicated in relation with the input event.
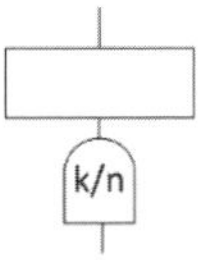 (k/n gate symbol)	k/n gate (k out of n)	The output event takes place when k of the n inputs take place.

Figure 12.4. *Connectors*

234   Risk Analysis

Symbol	Name	Significance
	Input symbol	The upper part of the tree is on the page containing the symbol 101.
	Output symbol	This event is used on another page.

Figure 12.5. *Transfer symbols*

12.4. Implementation of the method

The methodology for building a fault tree is shown in Figure 12.6. Stage 1 consists of choosing a top event that we will explain later. This is generally provided by the previous PHA.

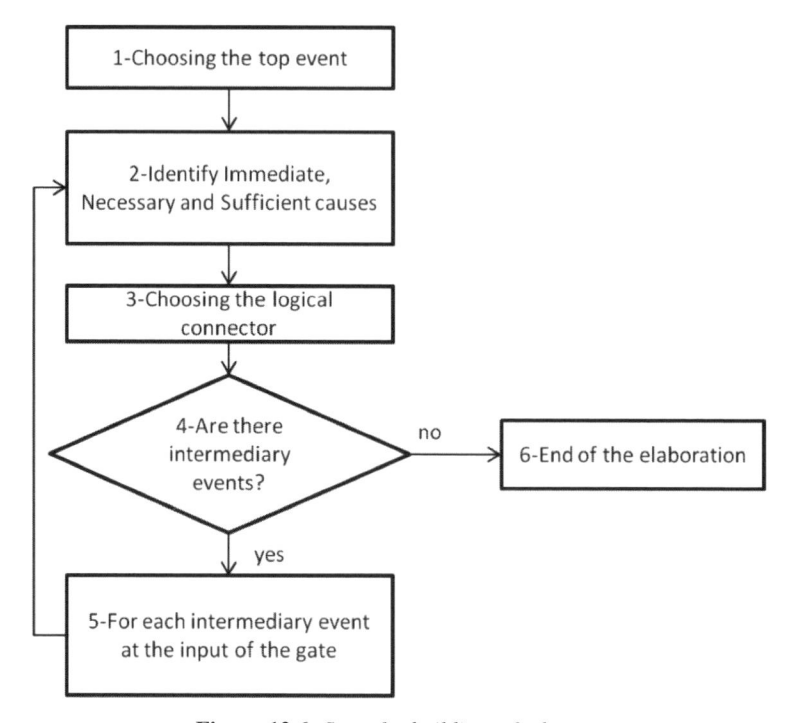

Figure 12.6. *Steps for building a fault tree*

Stage 2 consists of looking for the INS causes of the event being studied. It is carried out by asking the following questions:

1) What is necessary for this event to take place?

2) Is the identified cause a direct cause?

3) Is it related to abnormal behavior? If not, it should be discarded.

4) Can it be considered as a primary (internal), secondary (external) or command (external) cause?

5) Is it a variable deviation or an undesirable event at the level of the system?

6) Is the identified cause sufficient to trigger the occurrence of the event that is being analyzed, either immediately or later upon the occurrence of a event occurring during normal operation? If not, we should keep looking for the other necessary cause, and go back to question 2.

At the end of this identification we can classify each cause as a basic event, an intermediary event or a non-developed event. Throughout the process of finding out the causes, we must adopt a conservative attitude. For example, if normal operation triggers an event, we will not automatically assume that a fault in the equipment will stop this event from occurring.

Stage 3 of Figure 12.6 defines the logical connector that connects the causes to the event being analyzed. In certain cases, this may need a reorganization of the INS causes with the introduction of intermediary events, in order to make the OR and AND connectors appear as appropriate.

For example, let us consider the case of an electrical kettle. If the lid is closed, it must automatically stop when the water is hot because of an internal contactor. If it does not stop, this could trigger an overheating of the kettle, then a short circuit, which, if the circuit breaker is faulty, could then lead to a fire of electrical origin. The fire therefore has the following causes:

– malfunctioning of the circuit breaker;

– forgetting to close the lid, or fault of the contactor.

We can therefore define two intermediary events, which are connected via an OR gate to the event "fire", whereas the causes of these intermediary events are connected via AND gates and each represents a set of sufficient causes (Figure 12.7).

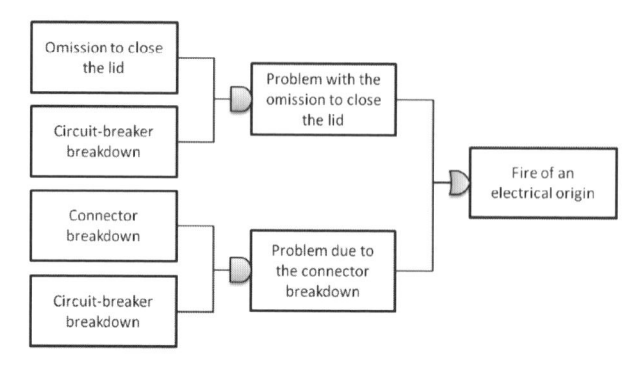

Figure 12.7. *Building in the case of several sets of sufficient causes*

Stage 4 consists of examining if there are causes classified as intermediary events. Should there not be any, the analysis is finished.

In stage 5, we select an intermediary event, and we repeat the procedure starting from stage 2 in order to study all of its causes. We always choose to explain one of the intermediary events that are closest to the top event, so as to homogeneously develop all of its branches.

To follow this procedure, we must observe the following two rules:

– Complete each gate: all the inputs of a gate must be defined before continuing with another gate.

– Do not connect the output of a gate directly to the input of another gate; instead, introduce a precisely defined intermediary element.

Most often, a fault tree is built after a risk analysis such as PHA and/or FMEA. The building of the tree therefore facilitates the organization of the existing information by highlighting the logical connections that are detailed between the events, as well as serving to complete the analysis. Figure 12.8 shows the connections between the different types of graph and the analysis methods:

Fault Tree Analysis 237

– A PHA approach allows us to identify the "undesirable events" that will be analyzed in more detail[1].

– A FMEA or HAZOP approach allows for a more detailed analysis of the dysfunctions: it allows us to obtain the list of basic events, as well as other faults or deviations that could be either internal events within the fault tree or a part of the consequence tree, in particular, for those regarding the barrier faults.

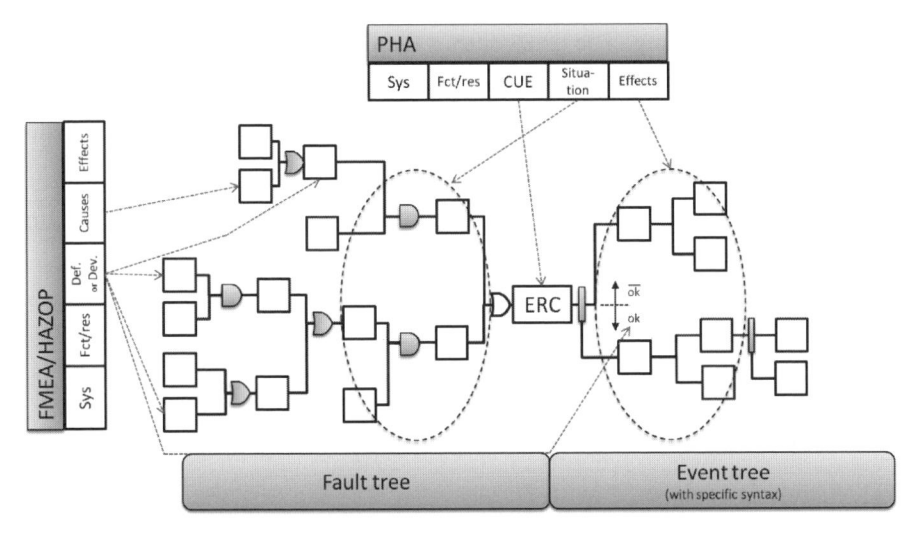

Figure 12.8. *Connections between the methods*

12.5. Qualitative and quantitative analysis

One of the major interests of the fault tree is to allow for the search for minimal cut sets, that is basic events combinations that will trigger the occurrence of the top event. These minimal cut sets can be used for calculating the probability of the top event, depending on the probabilities or the failure rate of basic events. They are also a means of qualitatively analyzing the tree, identifying the faults whose impact is the most significant. Indeed, a cut set with few elements indicates a significant vulnerability that is due to one or several faults, whereas a cut set with a significant number of

---

1 The causes and the effects noted in the table refer to events that are not always the immediate causes and effects. This helps to build a synthetic view.

events indicates a lower vulnerability, since all the faults must take place at the same time in order to trigger the occurrence of the top event.

12.5.1. *MOCUS algorithm*

The method for obtaining a cut sets (MOCUS) algorithm allows us to determine all the minimal cut sets of a fault tree. It is a *top-down* algorithm, which consists of replacing each gate starting from the top and generating the cut sets as it goes along. The procedure is the following:

– Writing the logical equation that allows us to calculate the top event on the basis of the basic events. The AND gates are noted with the symbol "." and the OR gates are noted with the symbol "+".

– Developing this equation in order to express the top event as a sum (disjunction) of products (conjunction):

$$S = T_1 + T_2 + \ldots + T_n$$

where each term is a product $T_i = E_{i1} \ldots E_{im}$.

– Simplifying this equation by using the following rules:

- rule 1: replace E^k with E;

- rule 2: let T_i and T_j be two terms of the obtained sum S. If T_i can be written as $T_i = T_j K$, then T_i is a redundant term, which can be removed from the sum.

Each term of the final expression of the sum is a minimal cut set.

To illustrate this approach, let us consider the tree shown in Figure 12.9. The intermediary events are noted from A to D and the basic events from 1 to 4. Let us write the logical expression of the top event A as:

$$A = B.D = (1.C).(2+4) = (1.(2+3))(2+4) = 1.2.2 + 1.2.4 + 1.3.2 + 1.3.4$$

– Applying rule 1 allows us to simplify $1.2.2 = 1.2$, which gives $A = 1.2 + 1.2.4 + 1.3.2 + 1.3.4$.

– The second rule allows us to eliminate the terms 2 and 3, which both contain the term 1.2.

Fault Tree Analysis 239

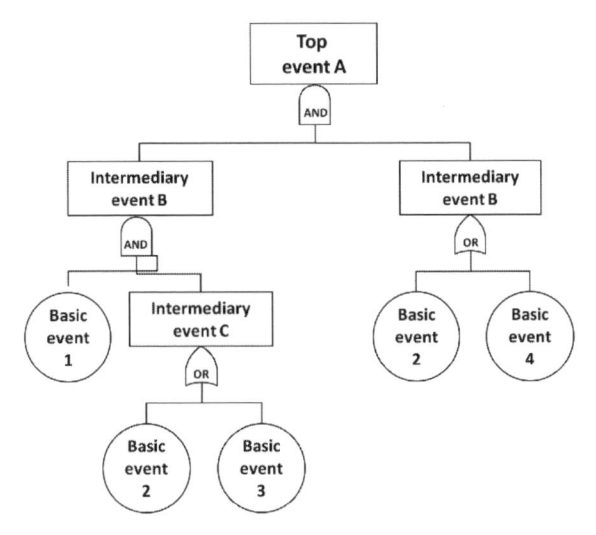

Figure 12.9. *Example for determining minimal cut sets*

We finally obtain:

$$A = 1.2 + 1.3.4$$

The two minimal cut sets are therefore $\{1, 2\}$ and $\{1, 3, 4\}$. Let us note that the event 1 is a cause that is common to the two cut sets.

12.5.2. *Probability calculations*

One of the interests of the fault trees is to allow the calculation of the probability of the top event. The probability of the occurrence of this event, noted as S, can be written as:

$$P_S = P(CS_1 + CS_2 + \ldots + CS_n)$$

where $\{CS_1, \ldots CS_n\}$ are the minimal cut sets of the tree. Let us note the probability of the cut CS_i with $P(CS_i)$. The Poincaré theorem allows us to write:

$$P_S = \sum_{i=1}^{n} P(CS_i) - \sum_{i<j} P(CS_i \cap CS_j) + \sum_{i<j<k} P(CS_i \cap CS_j \cap CS_k)$$
$$- \ldots + (-1)^{n+1} P(CS_1 \cap \ldots \cap CS_n)$$

Starting from this expression, we can obtain a lower and upper bound for P_s:

$$\sum_{i=1}^{n} P(CS_i) - \sum_{i<j} P(CS_i \cap CS_j) \leq P_S \leq \sum_{i=1}^{n} P(CS_i) \qquad [12.1]$$

The probability P_s can be approximated using the following relation, called approximation for rare events:

$$P_S \simeq \sum_{i=1}^{n} P(CS_i)$$

It can be used when the simultaneous probabilities of occurrence of the two minimal cut sets are low, compared to the probability of these cut sets. This is the case if the cut sets are independent and the probabilities are small. If this is not the case, this approximation is conservative and represents an upper bound of the probability of the top event.

Moreover, if the events of each minimal cut set are independent, the probability of a cut sets is equal to the product of the probabilities of each event:

$$P(CS_i) = \prod_{i=1}^{m} P(E_i)$$

In case a basic event E represents a fault whose fault rate λ is known and constant, we can deduce the occurrence probability for a period T with:

$$P(E) = 1 - e^{-\lambda T}$$

Let us remember that, in this case, the fault rate can be obtained from the mean time between failures (MTBF) $\lambda = 1/\text{MTBF}$ (see Appendix 4).

Let us take once again the example in Figure 12.9. Let us suppose that the failure rates are known and constant:

$$\lambda_1 = 10^{-5} h^{-1}, \; \lambda_2 = \lambda_3 = 10^{-4} h^{-1} \text{ and } \lambda_4 = 5 \times 10^{-4} h^{-1}$$

Fault Tree Analysis 241

The occurrence probabilities in 1 year of basic events are $P_1 = 0.0036$, $P_2 = P_3 = 0.0359$ and $P_4 = 0.0181$, and that of the top event is $P_A \simeq 0.00013$. The relation established above [12.1] allows us to give the following bounds for the probability of the top event $0.00013312770 \leq P_A \leq 0.00013312801$. The rare events approximation that takes into account the upper limit is very good because the two bounds only differ at the ninth decimal.

12.5.3. *Importance measures*

It might be interesting to be able to tell the importance of a particular event in the occurrence of the top event. Several approaches were proposed for assessing this importance.

The first approach, called "cut set importance", consists of assessing the contribution of each minimal cut set to the probability of the top event:

$$I(CS) = \frac{P(CS)}{P_S}$$

This measure allows us to classify the impact of each minimal cut set. In our example, $I(\{1,2\}) = 0.98$ and $I(\{1,3,4\}) = 0.02$.

The second approach, called the Fussell–Vesely importance factor, consists of evaluating the contribution of each event E to the probability of the top event, noted S. It is defined by the conditional probability of E knowing S:

$$I(E) = P(E|S) = \frac{P(E \cap S)}{P_S}$$

A good approximation of $I(E)$ is obtained by summing up the probability of all the minimal cut sets containing E and dividing this sum by the probability of the top event:

$$I(E) = \frac{\sum_{CS:E \in CS} P(CS)}{P_S}$$

In the case of our example, the event 1 has an importance of 1, the event 2 has an importance of 0.98 and the events 3 and 4 have an importance of 0.02.

This factor allows us to organize into a hierarchy the actions that must be carried out in order to improve the reliability of the system. In our example, it is clear that the event 1 must be treated with priority.

The third approach, proposed by Birnbaum [BIR 69], defines the importance of an event as the sensitivity of the probability of the top event to the probability of the considered event:

$$I_B(E) = \frac{\partial P_s}{\partial P_E}$$

We can show that this size is equal to the difference of the probability of the top event when E takes place and of the probability of the top event when E does not take place. We calculate P_s by first considering $P_E = 1$, then $P_E = 0$.

In our example, we have $I_B(1) = 1 - 0 = 1$, $I_B(2) = 0.004$, $I_B(3) = 7\,10^{-5}$ and $I_B(4) = 1.3 \times 10^{-4}$.

This Birnbaum importance factor allows us to determine the events that must be treated with priority and, during the probabilist assessment phase, it allows us to determine the events for which the reliability data collection must be done most carefully.

12.6. Connection with the reliability diagram

A reliability diagram or "success diagram" consists of representing a system by blocks generally corresponding to components, functions or sub-systems. The diagram has an input and an output, and each block can be seen as a turned off switch if the block is functioning well and is otherwise switched on. The function will be implemented if there is at least one pathway between the input and the output. If all the components are necessary, we have a series diagram. If certain components are redundant, they are parallel (Figure 12.10).

A fault tree using only AND and OR gates can always be converted into a reliability diagram. An AND gate generates a set of blocks in parallel, since all the elements must be failing in order to trigger the occurrence of the resulting event. An OR gate is converted into a series diagram. In the case of the example in Figure 12.9, we obtain the diagram presented in Figure 12.11. This block diagram representation is described in the IEC 61708 norm and is used to calculate the reliability of non-repairable complex systems (see Appendix 4).

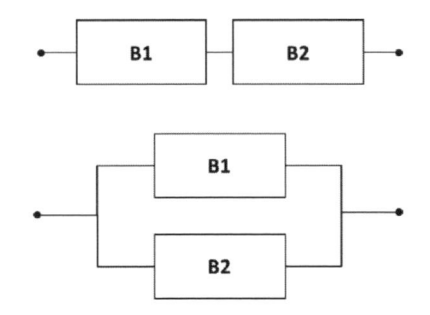

Figure 12.10. *Series and parallel diagrams*

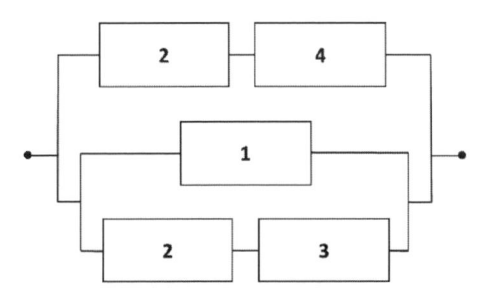

Figure 12.11. *Example of a reliability diagram*

12.7. Model-based approach

The use of a model-based approach for the risk analysis facilitates the building of the fault tree. Indeed, the PHA and FMEA analyses that are carried out using this approach are represented internally as an event graph that contains the causality connections between all the events created during the analysis. Consequently, the structure of the fault tree can be generated easily. We need only to specify the logical connections between the events because they are not defined in the analysis tables of the different methods.

An example of a tree built starting from FMEA is given in section 12.8.2.

12.8. Examples

12.8.1. *Desk lamp*

Starting from the FMEA analysis discussed in Chapter 9, it is possible to build the fault tree of the electric shock, as shown in Figure 12.12. This can be built automatically, except for the information on the type of connector which is in the FMEA table.

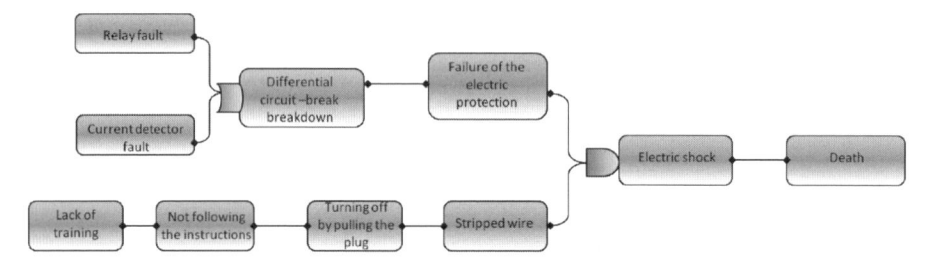

Figure 12.12. *Fault tree for the example of the desk lamp*

12.8.2. *Chemical process*

12.8.2.1. *Fault tree of the pumping function*

As we have indicated, the building of a fault tree is generally done after a PHA and/or FMEA analysis is carried out. This previous analysis allows us to structure the building of the tree and to define the level of detail for searching for the causes because it already contains the majority of useful events. If the analysis is carried out using a model-based approach (section 12.7), then the building of the tree is automatic.

Let us consider the example shown in Figure A7.2. The FMEA is given in Figure 9.16. The corresponding fault tree is shown in Figure 12.13. It was automatically generated starting from the FMEA analysis (Figure 12.13). Adding the information regarding pump redundancy and therefore modifying the input connector of the event "mechanical breakdown on two pumps" was sufficient for reaching the final tree.

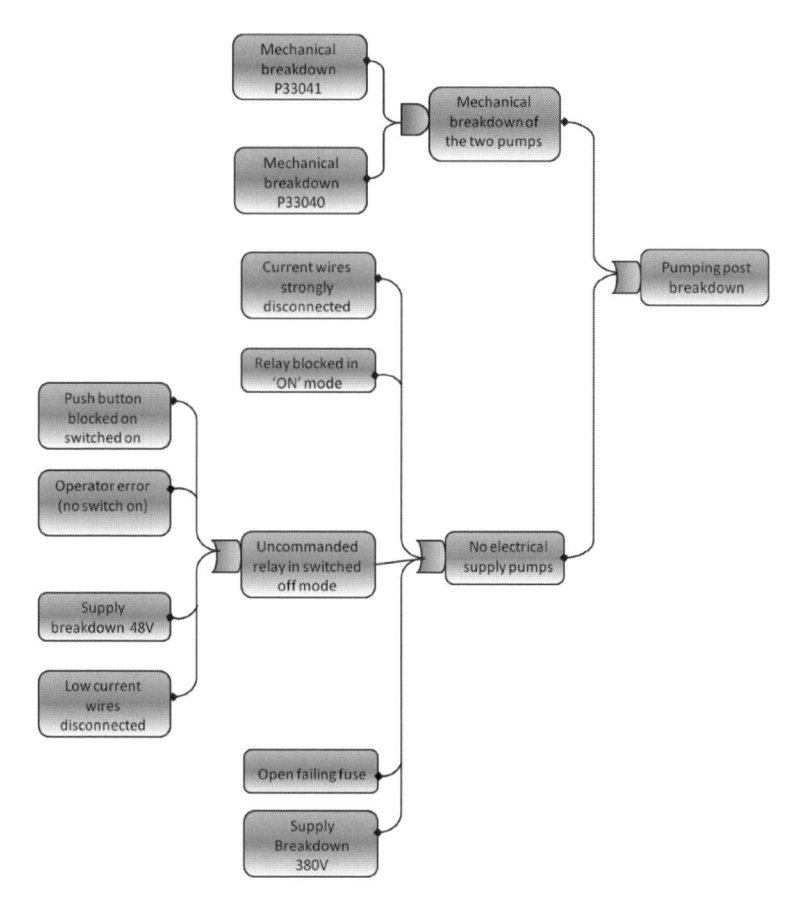

Figure 12.13. *Fault tree generated from the FMEA for the example of the pumping post*

12.8.2.2. *Fault tree of the explosion*

One of the dangerous events having the most serious consequences is explosion. It is therefore essential to assess the probability of this explosion taking a quantitative approach and therefore to build the fault tree leading to this explosion. This tree is shown in Figure 12.14. It was built starting from the PHA analysis and the FMEA analysis of the chemical process. As previously, the input connector of the explosion was specified and defined as an AND gate.

246   Risk Analysis

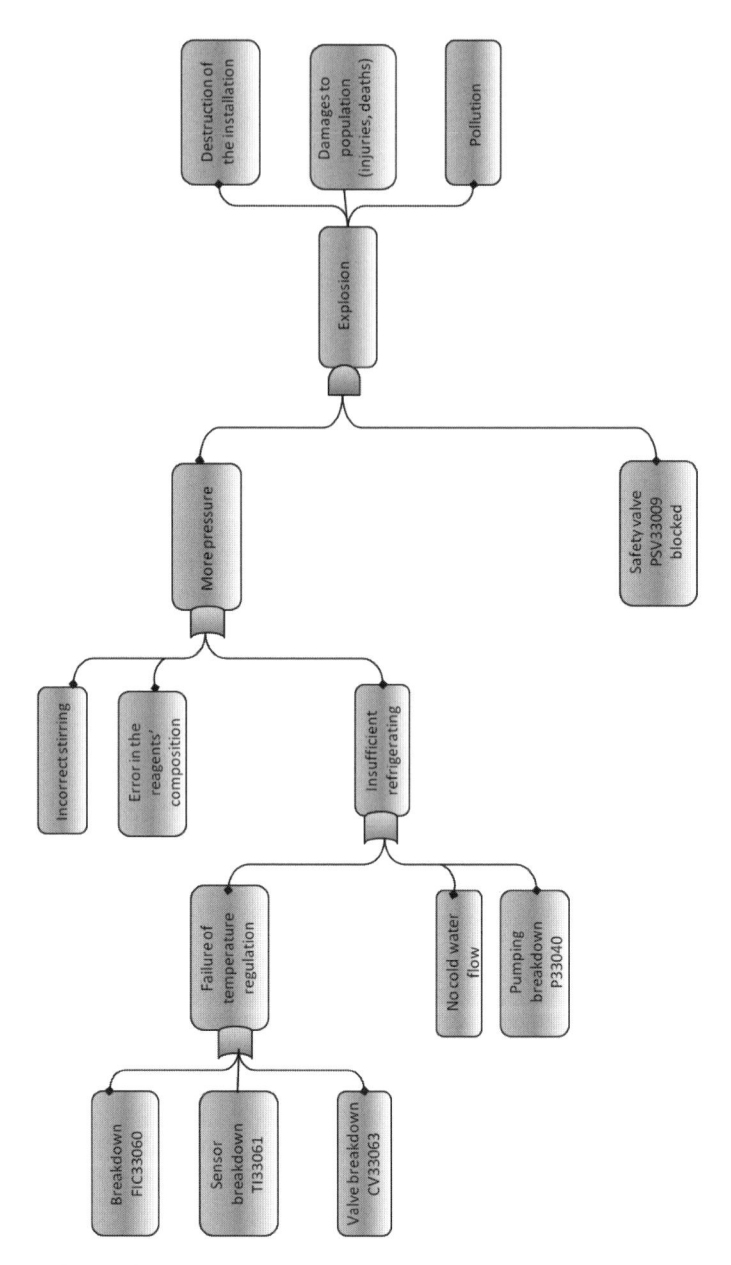

Figure 12.14. *Fault tree generated from the FMEA for the explosion*

12.9. Common cause failure analysis

12.9.1. *Introduction*

A common cause failure (CCF) is a fault whose origin has the same root cause as an entire set of causes. These are not independent. The CCF analysis is a very important aspect that must be considered in risk analyses because CCF significantly increase the vulnerability of a system.

EXAMPLE 12.1.– In the case of the pumping unit, the mechanical breakdowns of the two pumps are assumed to be independent. We can imagine that these breakdowns can be due to a manufacturing error, if they come from the same provider or due to a maintenance error. In this case, they are no longer independent, and they have the same common cause.

When such a dependence exists between the failures, the independence hypotheses between the events used for probability calculation are no longer valid. The common cause must be considered.

This can be modeled either explicitly or implicitly. In the case of an explicit modeling, we will, for example, have a modification of the failure tree in order to highlight the maintenance errors (Figure 12.15). The minimal cut sets are modified, a new cut sets appears along with the common cause and the probability of the top event is modified as a consequence.

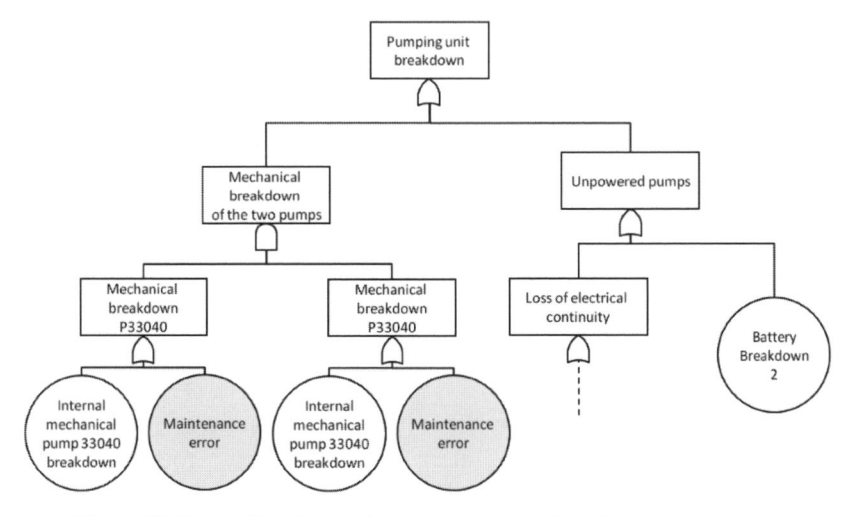

Figure 12.15. *Modification of the tree so as to consider the common causes*

Another approach is to implicitly model the dependencies between the failures by correcting the probability calculation of the failures of redundant systems. The method of the β-factor is the method used most often. We will detail this in the following.

12.9.2. *Identification of common causes*

To identify failures due to common causes, we search for possible dependencies between several failures. We distinguish between the following [NUR 07]:

– Internal dependencies: these are the causes connected to internal phenomena that affect several elements at the same time, such as, for example, the fault of an element due to overheating, which also impacts all the other elements of the system.

– External dependencies: these are due to a factor that is external to the system analyzed. There are two main categories:

- those due to physical phenomena, such as vibrations, or an earthquake, which affect several elements;

- those related to human actions, such as a maintenance error.

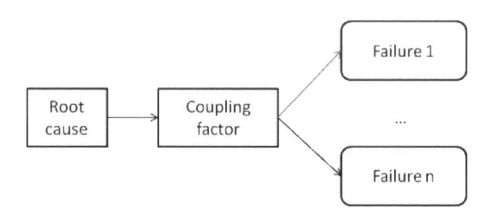

Figure 12.16. *Two stages for commun causes identifications*

The study of common causes can be carried out in two stages (Figure 12.16) [NUR 07]:

– Identifying the root cause, which provides the initiating event:

- the events connected to the external environment, such as natural hazards (flooding, earthquake, etc.), fire, humidity, too high or low temperatures, electromagnetic fields, radiations, bacterial contamination, dust and animal actions (rodents);

- human errors: unintentional actions, inadequate procedures, not following the procedures, lack of practice or training;

- faults connected to the design or implementation of the hardware or software;

- events internal to the system, related to component wear, manufacturing faults and interaction problems between elements.

– Studying the coupling factor:

- an identical design;

- an identical hardware;

- an identical software;

- the same exploitation staff;

- the same service or maintenance staff;

- the same procedures;

- the same localization;

- the same physical environment.

An action upon these coupling factors is often the best means of limiting the occurrence of the coupled faults. Let us cite, for example, the design of the same software by two different teams or the different locations of two data storage centers.

12.9.3. *Common cause analysis*

Let us consider two redundant elements of a system, the fault of each being noted with FM_1 and FM_2, respectively. We can write:

$$P(FM_1 \cap FM_2) = P(FM_1|FM_2).P(FM_2)$$

When the events FM_1 and FM_2 are independent, we obtain:

$$P(FM_1 \cap FM_2) = P(FM_1).P(FM_2)$$

When this expression is not true, the events are said to be dependant. Let us consider a random event noted by C, which represents the occurrence of a cause common to FM_1 and FM_2. Let us suppose that the probability of the occurrence of these modes divided by the occurrence of the common cause is low in relation to the probability of the occurrence of the mode FM; let us also assume that the probabilities are small, which is often the case with classical situations. We write:

$$P_s = P(FM_1 \cap FM_2 \cap C) + P(FM_1 \cap FM_2 \cap \overline{C})$$
$$P_s \simeq P(C) + P(FM_1).P(FM_2).(1 - P(C))$$

which corresponds to the fact that the system is failing if the common cause is true or, if it is not true, to the fact that the two systems are failing at the same time.

EXAMPLE 12.2.– Let us take $P(FM_1) = 10^{-3}$ and $P(FM_2) = 10^{-3}$ as numerical values, with a common probability cause $P(C) = 10^{-5}$. The probability of failure of the global system goes beyond 10^{-6} without a common cause, to 10^{-5} with a common cause, although the latter is 100 times less probable. Taking into account common failure causes is therefore a very important point that must be considered in risk analyses.

12.9.4. *The β-factor method*

One of the simplest methods that is often used to model a common cause is called the β-factor [FLE 75]. To intuitively explain this principle, let us consider a system with n identical, redundant elements allowing for the execution of a given function. They can be represented by n elements in parallel on the reliability diagram. The main idea is to divide the failure rate of a component into two parts:

– the part corresponding to the common cause $\lambda^C = \beta\lambda$;
– the normal part noted by $\lambda^N = (1 - \beta)\lambda$.

The β-factor appears as the relation between the failure rate due to the common cause and the global failure rate. The estimation of the β-factor is a significant aspect of the method. In general, this factor has a value between

0.005 and 0.15 for the hardware faults and between 0.05 and 0.6 for the software faults.

The norm IEC 61508 [IEC 10b] gives an approach for determining this factor. It is based on a series of questions, each question being associated with a score. Using the obtained total, we can determine the β-factor with the table given in Figure 12.17.

Score	Logical electronic based system	Sensor or actuator
120 or more	0.5 %	1 %
70 to 120	1 %	2 %
45 to 70	2 %	5 %
Less than 45	5 %	10 %

Figure 12.17. *Determination of the beta factor*

EXAMPLE 12.3.– Let us take the example of a safety pressure measuring sensor that has a dangerous failure rate of $\lambda = 0.00001\,h^{-1}$. Two sensors are used in a safety instrumented system (SIS) so that in the case of overpressure, the SIS can stop the procedure. The sensors are used in parallel, but they have a common failure cause estimated with the help of a $\beta = 0.1$ factor. Let us note with P_a the probability of failure of the sensor. Let us calculate the probability of failure of the system at the end of a year, first without considering the common cause. We write $P_a = 1 - e^{-\lambda T}$ and $P_s = P_a^2$, which gives:

$$P_a = 1 - e^{-0.00001 \times 8760} = 0.083873 \text{ and } P_s = P_a^2 = 0.007305$$

Let us calculate the same probability by taking into account the common cause. Let us note with P_c the probability of this common cause. By using the β-factor, we have $P_c = 1 - e^{-\beta \lambda T}$, $P_a = 1 - e^{-(1-\beta)\lambda T}$ and $P_s = P_a^2 + P_c - P_a^2 P_c$, which gives:

$$P_c = 1 - e^{-0.000001 \times 8760} = 0.008721743,$$

$$P_a = 1 - e^{-0.000009 \times 8760} = 0.0758123 \text{ and } P_s = 0.01441911.$$

The probability of failure has been multiplied by 2.

Chapter 13

Event Tree and Bow-Tie Diagram Analysis

13.1. Event tree

13.1.1. *Description*

The "consequence tree" or "event tree" method is an approach that allows us to identify and represent the sequence of events that take place as a consequence of a dangerous event, called the initiating event of the tree. The main interest of the method is that it allows us to consider the different alternatives that can take place after each event. The diagram obtained has a tree-like graphical structure (Figure 13.1), where the root corresponds to the dangerous initiating event and the terminal elements represent the different final alternatives that may take place. A quantitative analysis of the tree allows us to assess the probabilities of these different alternatives.

As we will show in the following, an event tree can be represented by a fault tree. However, when we are concerned with the evolution of a system after the occurrence of a dangerous event and we also wish to consider the safety barriers and/or decisions of the operator, the representation through an event tree becomes clearer. The first examples of this representation can be found in the safety study of the nuclear reactor [NUR 75], for which the team has developed the event tree approach in order to compress the fault tree representation, which seems more difficult to grasp. This method is used in the risk assessment phase (Figure 13.2), generally when the safety barriers are being performed and reviewed.

254   Risk Analysis

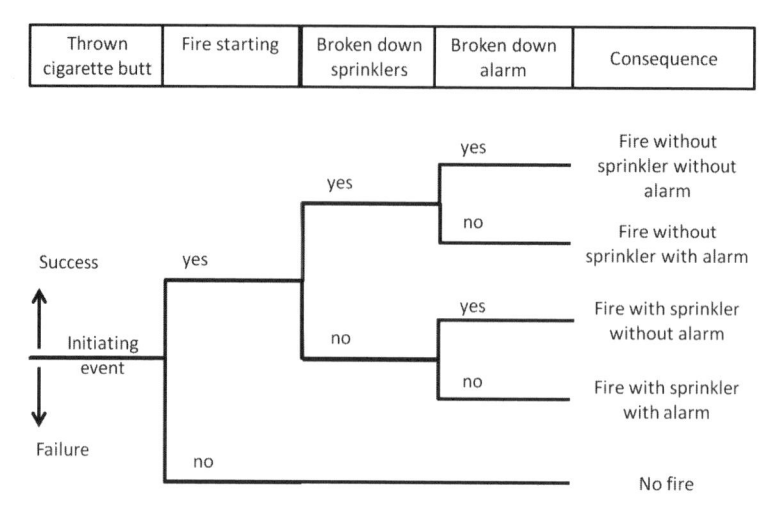

Figure 13.1. *Example of an event tree*

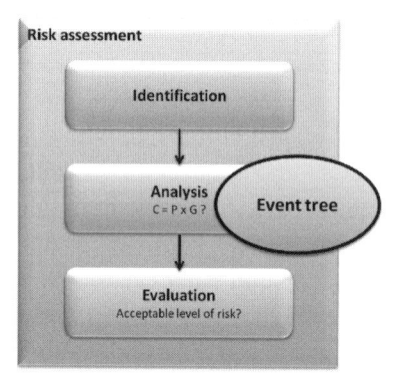

Figure 13.2. *The event tree in the risk assessment process*

REMARK 13.1.– The event tree is used for representing the operating actions face to a dangerous event under the name of operator action event tree (OAET) (section 7.7.3).

13.1.2. *Building the event tree*

An event tree can be seen as a decision tree that is built on the basis of a sequence of events, going from the initiating event to the final consequences, and through a certain number of pivotal events (Figure 13.3). Each of these events may or may not take place, leading up to a binary decision tree. These events generally represent:

- the occurrence of a phenomenon;
- the fault of a system element, i.e. often a barrier;
- an operator's decision.

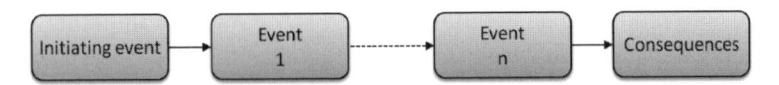

Figure 13.3. *Events sequence describing a scenario*

The event tree is obtained by explaining the different paths that lead from the initiating event to the consequences (Figure 13.4). There are event trees with decisions that are not binary and therefore, have more than two branches following each pivotal event. We rarely find them and they can be transformed into a binary tree by introducing intermediary pivotal events.

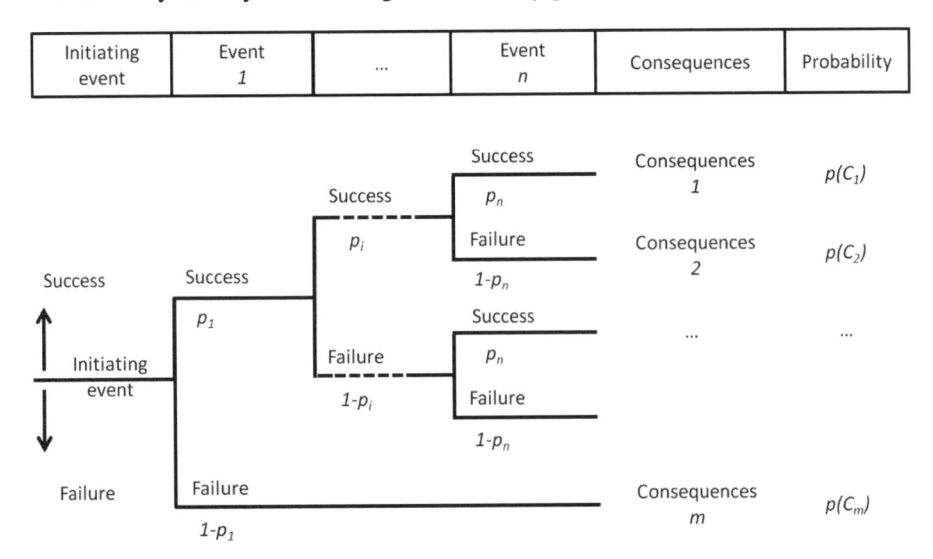

Figure 13.4. *Event tree obtained from the scenario in Figure 13.3*

The starting point for building an event tree is the identification of a dangerous event, which will be the initiating event for the tree. The fault tree analysis aims to analyze, via a deductive process, the causes of a dangerous event. Conversely, the event tree analysis starts from a dangerous event and seeks to determine all the consequences that flow from it.

The system and context definition phase and the dangerous events identification phase are performed before the construction of an event tree. Oftentimes, the latter is performed after a PHA analysis has been carried out to analyze the significant scenarios. The initiating event of the tree is chosen from the dangerous events identified in the PHA. Once the initiating event is determined, the following stages are:

1) identifying pivotal events, and final consequences;

2) building the basic scenario;

3) building the tree on the basis of the scenario;

4) analyzing the probabilities of the different events if a quantitative analysis is needed;

5) describing the different alternatives and calculating the respective probabilities;

6) proposition of prevention measures if required.

In general, the identification of pivotal events is done by analyzing the safety functions that will act in response to the initiating event. These can be classified into the following categories:

– detection functions of the initiating event;

– alarm functions;

– limitation functions with the aim of stopping the initiating event;

– mitigation functions, which aim to reduce the significance of the effects of the initiating event.

These functions can be ensured by passive devices, such as a protective cover, or active devices, such as automatic safety systems or even operators that act according to some pre defined procedures. Chapter 15 proposes an approach for assessing the performance of these different types of devices.

EXAMPLE 13.1.– Let us consider a chemical reactor in which an exothermic reaction occurs, as described in Appendix 7. To respond to a failure of the

Event Tree and Bow-Tie Diagram Analysis 257

refrigeration system, the following functions have been designed:

– redundant detection of the rise in temperature (TI33071);

– alarm aimed at the operator;

– refrigeration and stopping of the reaction via the sprinkler system.

These different functions will allow us to build the event tree presented in Figure 13.5.

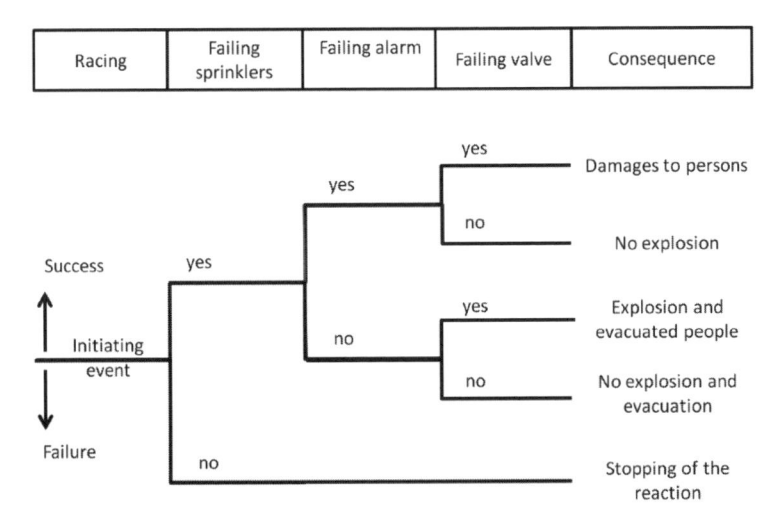

Figure 13.5. *Example of the construction of an event tree*

13.1.3. *Conversion into a fault tree*

An event tree representing the different possible sequences can be transformed into a fault tree in introducing the pivotal events and the associated events that represent their negation. Each branch is transformed into an AND gate between the event preceding the branch, and either the pivotal event, or its complement. Figure 13.6 shows an example of such a conversion, aiming to obtain a fault tree that leads up to the final alternative, where all the pivotal events occur. The representation of all the alternatives can quickly lead to a tree that is difficult to understand.

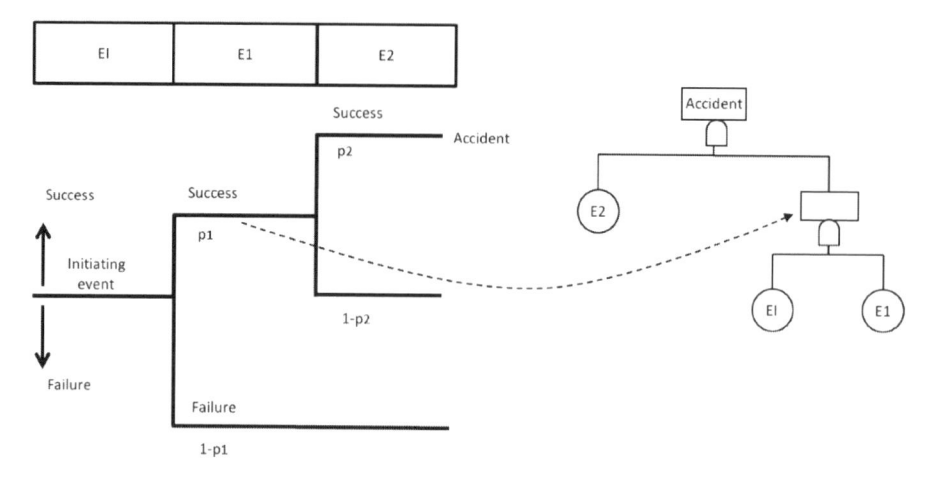

Figure 13.6. *Conversion of an event tree into a fault tree*

13.1.4. *Probability assessment*

The event tree can be used to calculate the probability of the different alternatives generated from the initiating event. To do this, we must know the probability of the initiating event, and the conditional probability of each pivotal event, depending on the sequence already implemented. The alternative obtained by considering all the success branches can be expressed as:

$$P(C_1) = P(E_{Init})P(E_1|E_{Init})P(E_2|E_1 \cap E_{Init})\ldots P(E_n|E_{n-1} \cap \ldots E_{Init})$$

Most often, the probability of each pivotal event does not depend on the path followed, which is the same as saying that all events are independent. Therefore, we can write:

$$P(C_1) = P(E_{Init})P(E_1)P(E_2)\ldots P(E_n)$$

When the failed alternative is taken, the probability will be $1 - P(E_k)$ instead of $P(E_k)$ because the tree is binary. We can also calculate the probabilities of all the alternatives (Figure 13.7).

Event Tree and Bow-Tie Diagram Analysis 259

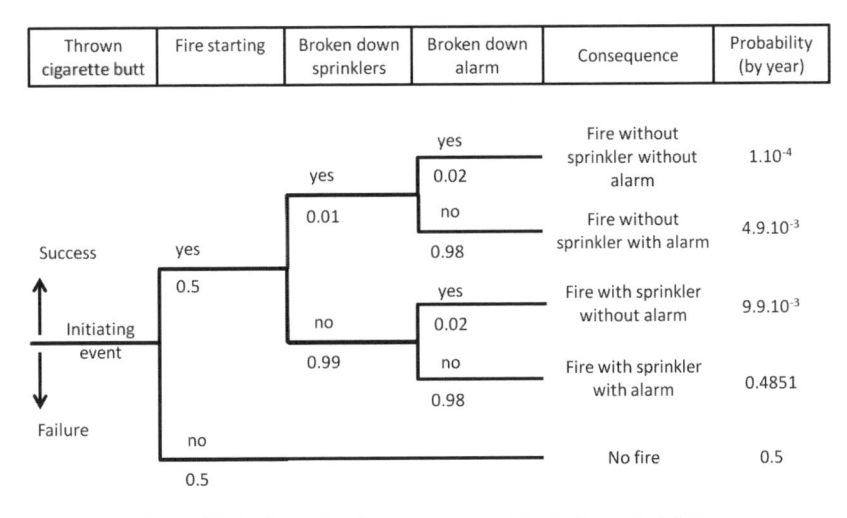

Thrown cigarette butt	Fire starting	Broken down sprinklers	Broken down alarm	Consequence	Probability (by year)
		yes 0.01	yes 0.02	Fire without sprinkler without alarm	1.10^{-4}
Success ↑ Initiating event ↓ Failure	yes 0.5		no 0.98	Fire without sprinkler with alarm	$4.9.10^{-3}$
		no 0.99	yes 0.02	Fire with sprinkler without alarm	$9.9.10^{-3}$
			no 0.98	Fire with sprinkler with alarm	0.4851
	no 0.5			No fire	0.5

Figure 13.7. *Example of an event tree with all the probabilities*

13.2. Bow-tie diagram

13.2.1. *Description*

A bow-tie diagram represents, in the same view, the relationships between a dangerous event, its causes and its consequences. It also allows us to represent the barriers that were implemented for reducing occurrence probability of the main event and those for reducing the occurrence probability of the consequences. This type of diagram was introduced in the 1990s by Shell, at the origin of this type of tool. It has since been used in different industrial sectors. This representation mode plays a central role in the ARAMIS method [SAL 06] developed within a European project at the beginning of the 2000.

This type of representation can be used as a qualitative or quantitative tool. In the former case, the representation is simplified, and it can globally highlight a set of causes and consequences for the dangerous event as well as the associated barriers (Figure 13.8). In the case of a more detailed analysis, the bow-tie diagram is a representation that combines a fault tree and an event tree with a slightly modified syntax.

In order for the representation of the event tree to be homogeneous with that of the fault tree, we must introduce the pivotal events in the bow-tie diagram, and not simply the different alternatives to the pivotal events. Moreover, considering that the alternatives are often related to the barrier failures, the alternatives are represented at the bottom of the graphical representation of the barriers:

– an edge coming out of the right side defines the path in case the barrier has a failure, i.e. the event that would have taken place if the barrier did not exist;

– an edge coming out below defines the sequence of events in case the barrier works as expected (Figure 13.9), i.e. an event that represents a mitigated consequence; this event is not always represented.

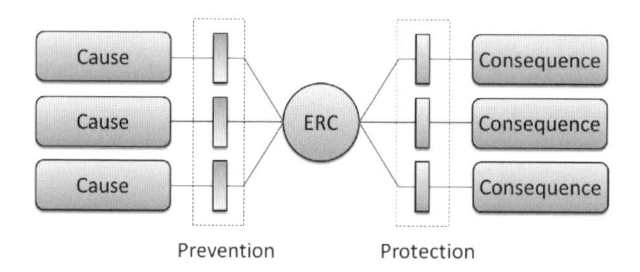

Figure 13.8. *Bow-tie diagram for qualitative analysis*

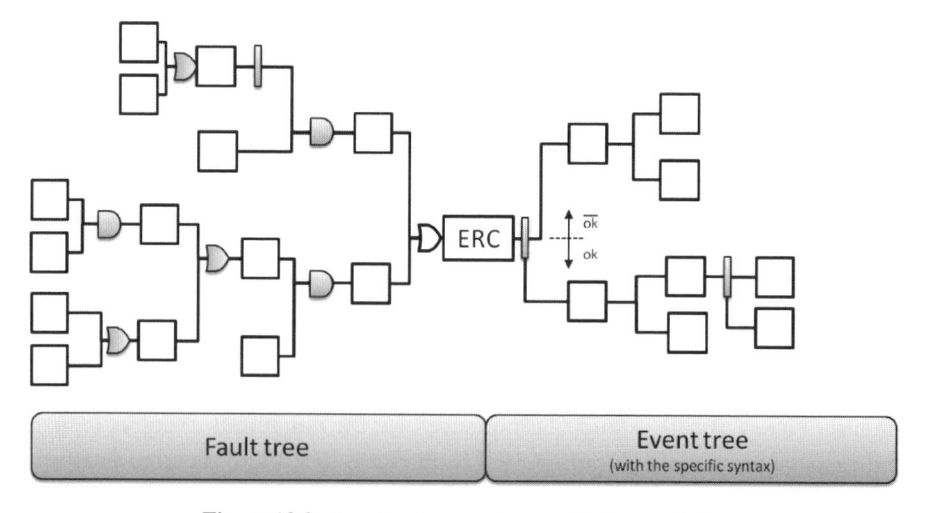

Figure 13.9. *Bow-tie diagram for quantitative analysis*

This representation is the one used by the INERIS [INE 06] and is similar to the one used in the barrier block diagram [SKL 06b] and in certain bow-tie representation softwares (Appendix 7). The diagram obtained groups together, in the same view, the causes of the Central Undesirable Event (CUE), its multiple consequences, represented by several connected edges at the right of an event, and the alternatives depending on the failures of the barriers.

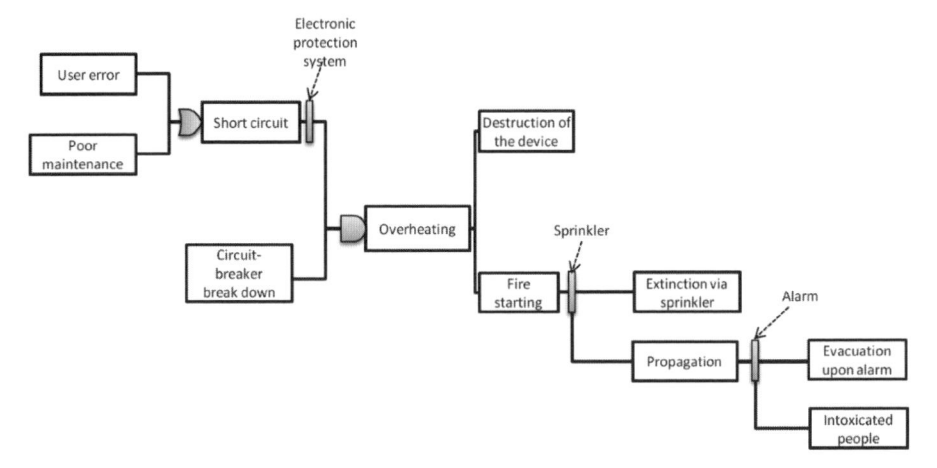

Figure 13.10. *Example of bow-tie diagram*

13.2.2. *Assessment of the probability*

The probability assessment in a bow-tie diagram can be done by applying the approach used for the fault tree and that used for the event tree. For the part corresponding to the event tree, we use the probability of failure on demand (PFD) of the barrier for the direct edge and 1-PFD for the edge in the case of normal functioning. This notion of "probability of failure on demand" and the calculation procedure are detailed in Chapter 15.

For the calculation to be correct, the events must be independent. It is, therefore, necessary to verify that there are no common causes for the failure of the barriers. However, if there are common causes, the independence hypothesis of the event tree alternatives' probability is no longer respected, which could lead to a significant underevaluation of the probability of damaging consequences.

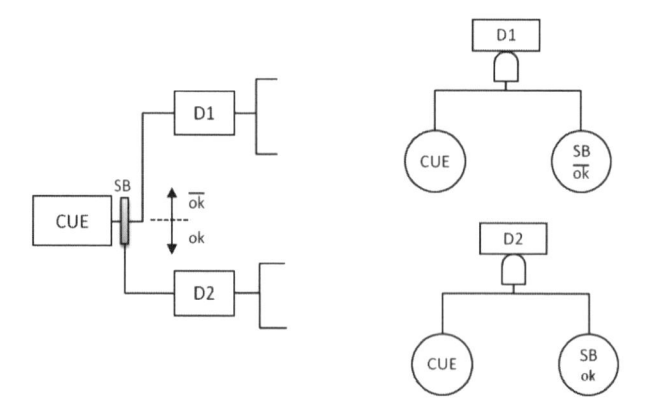

Figure 13.11. *Conversion of a bow-tie diagram into a failure tree*

13.2.3. *Conversion into a fault tree*

A bow-tie diagram can, just like an event tree, be converted into a fault tree. A barrier is transformed into an AND port and has the event that precedes the barrier, as well as the event "barrier failure" as input events.

REMARK 13.2.– The description of the barriers of an installation, in particular when they exist, can be carried out either with a vertical bar on the edge or by explicitly representing an event describing the failure of the barrier. It is necessary to choose one mode or the other, while being careful not to represent them twice as this might lead to erroneous probability assessments.

Chapter 14

Human Reliability Analysis

14.1. Introduction

14.1.1. *Objectives and context*

Very often, the system considered in a risk analysis includes interactions with one or several human operators, whether it is in the operation phase, the maintenance phase or, much earlier, in the design phase. Consequently, the origin of a dangerous event leading up to damages can often be found in what we call "human error", a term that we will define precisely in the following; we can understand it intuitively as an inadequate interaction between a human operator and the rest of the system.

In order to take into account the human factor, the need for analysis methods was quickly felt, particularly in the case of probabilistic risk analyses. When carrying out such an analysis, a model describing the succession of events that generate damage is built, using, for example, a fault tree. Among the primary causes we often find a certain number of events that characterize a human error. It is necessary to determine the occurrence probability of these events if we wish to undertake a quantitative study. Such events can also be found after the main dangerous event, when they are connected to safety barriers or other damage-amplifying phenomena.

The need for an approach capable of providing a failure probability for the human component has led to the development of numerous methods since the

1960s, with a strong increase in the production of methods taking place in the 1980s[1]. Therefore, the first methods developed, or their first versions, treated humans as mere components, carrying out actions characterized by failure modes for which an occurrence probability could be determined. Gradually, this viewpoint has evolved, and approaches have been proposed for taking into account the specificity of human operators.

The characteristics of a human being are indeed very different from those of a technical component. They achieve higher in certain situations and lower in others. For example, they are capable of adapting to a new situation and reacting in unpredictable situations. They can also quickly recognize forms, objects and situations by using their five senses and their intelligence. They can detect errors, including their own, and rectify them. However, for a certain number of tasks they are less successful than a technical system. This is, for example, the case of repetitive actions, or tasks of monitoring and scanning for long durations. The attention span of a human being is limited, and it fluctuates with time, depending on the environment and the stimuli received. Moreover, the characteristics of the human brain can be a risk factor, because of:

– its tendency to simplify in order to minimize the workload and facilitate the task;

– its tendency to retain the first possible explanation when explaining a situation;

– its mono-tasking functioning entails a fixation on the current task;

– a behavior that varies and even hesitates, which in turn triggers different decisions when faced with the same facts;

– excessive confidence in other personnel, written procedures and automatic systems.

The methods for assessing human reliability have, therefore, been developed in order to consider the human being with a specific approach within the task analysis of a socio-technical system, and to evaluate its contribution to the emergence of the risk, or to the opposite, to its reduction.

---

[1] The Three-Mile Island accident in 1979 is undoubtedly one of the reasons for this development.

14.1.2. *Definitions*

We all know the saying "to err is human". To be mistaken is therefore a property of man. However, mistakes are not synonymous with faults, and we should not assimilate the two. Moreover, in current language, the term "error" is usually used for statements that turn out to be quite vague; it is not used for inappropriate actions or behaviors.

Let us, therefore, analyze the notion of "human error" within the framework of risk analysis. The first definition was proposed by Swain and Guttmann [SWA 83].

DEFINITION 14.1.– *A human error is defined as an action that goes beyond the acceptable limits, those being defined by the system.*

With this definition, the human error is seen as a deviation from a normal or expected performance, considering the state of the system. A different definition was proposed by Reason [REA 90].

DEFINITION 14.2.– *Human error is a generic term that covers all the cases where a planned sequence of mental or physical activities does not achieve its desired objective, and when these failures cannot be attributed to chance.*

These definitions allow us, in theory, to characterize human error clearly. However, according to Hollnagel [HOL 93], this term is ambiguous and Hollnagel prefers the term "erroneous action". The problem is that the term "error" can be understood in different ways. It can, first, be used to designate the cause of an event leading up to the accident. It can also designate a failures of the cognitive process in planning the action or the sequence of actions, a failures in the implementation of the action, or even the omission to implement an action.

Moreover, even if the notion of human error is defined precisely, this term is often used to characterize the cause of an observed result, when the cause is related to a characteristic of human performance [WOO 94]. The term "erroneous action" allows us to eliminate these ambiguities:

DEFINITION 14.3.– *An erroneous action can be defined as an action that fails to produce the expected result and/or produces undesirable consequences.*

The definition of human reliability was introduced by Swain and Guttmann [SWA 83].

DEFINITION 14.4.– *Human reliability is defined as the probability of a person to:*

– *correctly perform an activity;*

– *perform no extraneous activity (with no connection with the task) that can degrade the system.*

DEFINITION 14.5.– *The probability of human error is defined by the relation between the number of errors taking place throughout the implementation of a task divided by the number of executed tasks that present us with opportunities for error:*

$$HEP = \frac{Number\ of\ errors}{Number\ of\ opportunities\ for\ error}$$

The number of opportunities for error is generally equal to the number of times that the task is carried out.

More than 35 methods of human reliability assessment have been published to date [BEL 09]. Very often, these methods are classified into at least two families.

The first-generation methods consider the operator as an entity that executes a sequential list of actions, which can be performed with performance outside acceptable limits. Potential failures are assumed to depend on the context in which the task is being carried out. These methods propose a characterization process of the human missions, referring to a set of generic tasks and/or factors that influence the performances, on the basis of which we can assess the probability of failing to perform that action. Depending on these methods, the probabilities can be obtained from tables or via expert judgments. Here are several examples of first-generation methods:

– Technique for Human Error Rate Prediction (THERP), 1963.

– Accident Sequence Evaluation Program (ASEP), 1987 (a simplified version of THERP).

– Tecnica Empirica Stima Errori Operatori (TESEO), 1980.

– Systematic Human Error Reduction and Prediction Approach (SHERPA), 1986.

– Success Likelihood Index Methodology, Multi-attribute Utility Decomposition (SLIM-MAUD), 1986.

– Human Error Assessment and Reduction Technique (HEART), 1986.

– Human Cognitive Reliability (HCR), 1984.

Second-generation methods better consider the cognitive processes. Currently, they need specific developments to be implemented in industrial applications [BEL 09], and could need quite significant resources. Here are some of the examples of these methods:

– Cognitive Reliability Error Analysis Method (CREAM), 1998.

– A Technique for Human Error Analysis (ATHEANA), 1996.

– Method for the Assessment of the Implementation of Safety Operator Missions (MAISOM), 1998.

We will detail some of these methods in the remaining chapter.

14.2. The stages of a probabilistic analysis of human reliability

The general procedure of a human reliability analysis (HRA) as defined by the IEEE STD 1082 standard is given in Figure 14.1. This figure also highlights the connections with the ISO 31000 standard. The main stages are the following:

1) Composing the team.

2) Getting to know the system that needs analyzing.

3) Building the system model by including the human aspects: the people in charge of the HRA work closely with the people building the system model in such a way as to include the role of the operator and their response to various dangerous events.

4) Analyzing the significant human interactions: the objective of this stage is to analyze the human tasks that could, potentially, have a significant impact in terms of risk.

5) Characterizing the tasks and their context in order to analyze the causes: the aim of this stage is to improve the model developed in stage 3 so as to be able to identify and analyze human errors.

6) Quantifying potential errors, by considering the context of the task.

7) Once all the scenarios related to human error have been assessed, the team studies whether any recovery actions are indeed possible. If they have not been considered in the initial model, then the model is updated.

8) A probabilistic risk analysis gathers various information regarding the installation in order to build the scenarios that represent the risks. To validate the results, an overview of the results by someone outside the team is necessary.

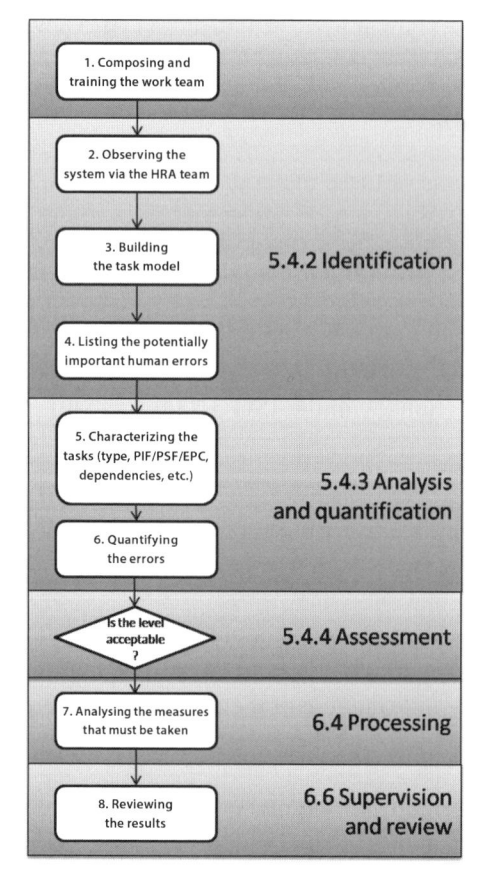

Figure 14.1. *The process of a human reliability analysis (IEEE STD 1082, 1997) and the ISO31000 procedure*

The last stage consists of writing up the documentation on the analysis. It provides a traceable description of the process used for the HRA, which is useful for the following stages of the risk management process and for quality assurance.

The methods proposed for implementing this HRA include three main stages (Figure 14.2):

– Task analysis:

   - model-based representation: a hierarchical task analysis (HTA) model, a flow model or an event tree, approaches that have all been discussed in section 7.7;

   - context analysis for identifying the factors influencing the performance, a process that is discussed in section 14.4.1.

– Identifying potential human errors by using a taxonomy that is appropriate for each method, and by relying on the classifications detailed in section 14.3.

– Quantification of errors with probability assessment, in either a qualitative or quantitative manner, and, depending on the methods, uncertainty assessment and a sensitivity analysis.

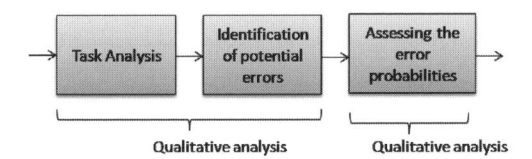

Figure 14.2. *The three stages of the analysis*

14.3. Human error classification

During an analysis of human reliability, the analyst must identify the possible errors for each action. To make the analysis as systematic as possible, the different methods propose different taxonomies of the possible errors, depending on the different characteristics of the task. They rely on generic classifications that we will present in the following.

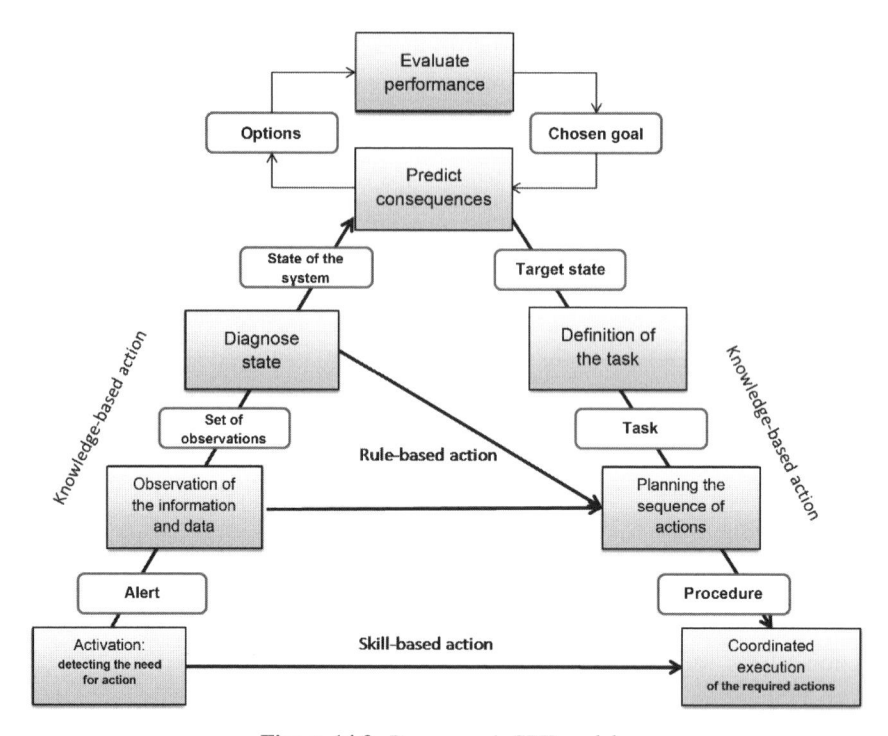

Figure 14.3. *Rasmussen's SRK model*

14.3.1. *Rasmussen's Skill – Rule – Knowledge (SRK) classification*

This classification (Figure 14.3) proposed by Rasmussen [RAS 83] in 1983 consists of classifying human activities into three categories, which will give way to different types of error. We distinguish between the following:

– The first category regards the activities leading to machine-like behavior (*skill-based behavior*). This is the behavior observed for automatic actions, which require very little or no mental effort at all. Examples of such activities are: writing, cycling, or to shift gears when driving a car. This is the case of an operator that carries out operations in a reflexive manner or operates upon their environment. The error probabilities generally used for this type of task range from $1/200$ to $1/20,000$.

– The second category regards the actions coming out of the procedural behavior (*rule based*), i.e. acting by following an explicit rule or procedure; this is the case where these actions are not mastered well enough to be applied mechanically. An example of such an activity is implementing an infrequent periodic test by an operator, or, in the case of a vehicle driver, adjusting a less-known option of his/her car by using the on-board computer. The error probabilities generally used for this type of task range from $1/20$ to $1/2,000$.

– The last category regards the actions ensuing from cognitive behavior (*knowledge based*), actions that need significant mental activity to solve a problem or make a decision. During such actions, the human being must first take the time to perceive the situation, then interpret the information and make a decision – which is not always simple – and finally implement the action(s). An example of this type of activity can be the action of a car driver who finds himself/herself in a queue of traffic and must choose whether to change the route or not. The probabilities generally used for this type of task range from $1/2$ to $1/200$.

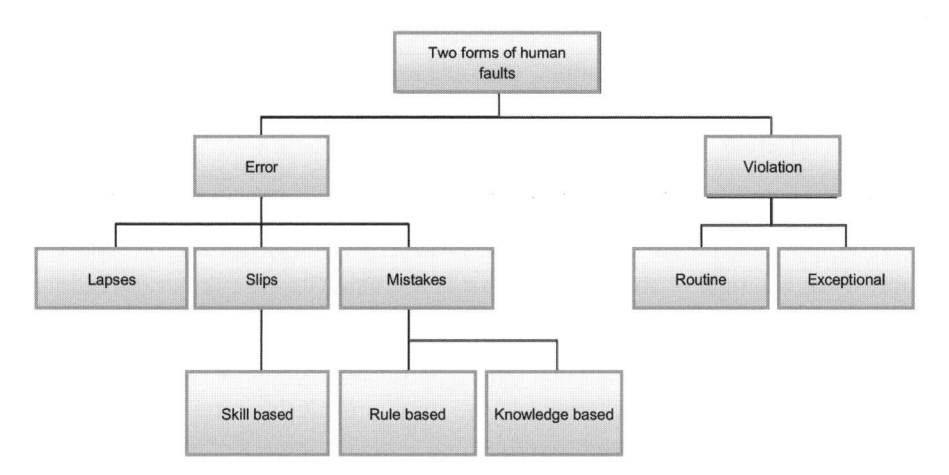

Figure 14.4. *The classification proposed by Reason [REA 90]*

14.3.2. *The Reason classification*

One of the points common to the definitions of human error discussed in section 14.1.2 is the notion of having the intention to obtain a certain result during the realization of an action by a human being. We can then distinguish between the errors at the level of intention and those at the level of realization of that action. The SRK classification was improved by Reason [REA 90] by taking into account this distinction. It defines four types of errors that could lead to an accident:

– The slips: these are the actions carried out with a correct intention, but that are incorrectly implemented. For example, the operator makes a mistake by closing the wrong valve because they confuse it for another valve.

– The lapses: these are the actions that are not carried out because of forgetfulness or a mere distraction. For example, the operator forgets to close the valve because he/she is interrupted.

– The mistakes: these are actions that are correctly implemented, but with an incorrect intention. The human being decides to implement an action that is not the right one. We can distinguish between:

- the mistakes due to a poor diagnosis, which leads to the application of an inadequate rule (*rule-based*). For example, the operator assesses that the reactor is in a correct state, relying on the temperature measurement he/she has taken, whereas in fact the temperature is above the threshold defined in the procedure;

- the mistakes due to a problem of insufficient cognitive capacity (*knowledge-based*). For example, the operator assumes that the cause of the rise in temperature is related to a refrigerating problem, whereas the actual cause is different.

– The violations: these are errors that are implemented deliberately, when a person applies a rule when he/she knows it is different from the one they should apply. An example of this type of error is the behavior of a driver exceeding the speed limit. We can classify violations into the following four categories:

- Routine violations: the behavior of the person is systematically opposed to the rule. An example is the choice made by certain pedestrians to cross outside of protected crossings.

- Exceptional violations: in a particular situation, for example in case of emergency, certain rules are not observed.

- Situation-specific violations: these violations are caused by the work environment, be it physical or organizational.

- Acts of sabotage, which are not normally considered by the HRAs.

14.3.3. *Errors of omission and commission*

Another classification was proposed by Swain and Guttmann [SWA 83]. It consists of distinguishing the errors according to the omission–commission dichotomy: a person can make an error, either because he/she does not do something required or he/she does not do it on time (omission) or because he/she does something incorrectly (commission). Therefore, the errors can be classified as:

– Omission error (intentional or unintentional):

- omission of a task;

- omission of a stage in a task.

– Commission error:

- selection error: a wrong action is carried out;

- execution error: an action is performed poorly or with wrong parameters;

- sequencing error: the stages are carried out in the wrong order;

- timing error: an action is carried out either too soon or too late.

This classification can be represented using a generic event tree and is well suited for quantitative risk analyses. It is used in the THERP method and the ASEP method [SWA 87].

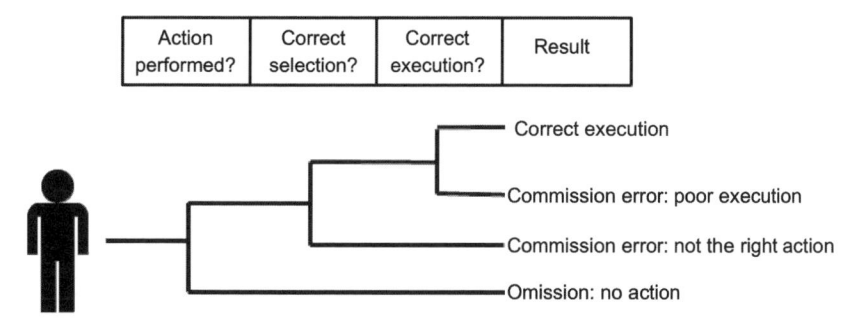

Figure 14.5. *Representation of the classification via an event tree*

14.3.4. *Pre-accidental and post-accidental errors*

The risk analyses that integrate the human factor distinguish between two types of errors, depending on whether they are related to the actual occurrence of the accident or to the implementation of a safety action. These are:

– pre-accidental human errors: these are the errors that can be placed at the origin of a dangerous event, which can be classified into:

274   Risk Analysis

- latent failures, such as test errors and maintenance errors. This is, for example, the case if we forget to re-validate the automatic removal from fire area system after a test;

- the failures directly triggering the dangerous event, which are operation errors, such as a bad connection, pressing a wrong button or non-compliance with an indicator prohibition;

– post-accidental human errors: these are the errors carried out after the occurrence of a dangerous event and that concern a recovery action or an action of implementing a safety barrier that is carried out in an incorrect fashion.

REMARK 14.1.– The human safety barrier assessment is the object of specific methods, such as the layer of protection analysis (LOPA) method (section 15.6) or the Omega 20 approach developed by INERIS [INE 09].

14.3.5. *Classification based on a cognitive model of the activity*

One last classification is carried out on the basis of the cognitive model of the activity. The CREAM method [HOL 98] defines the notions of phenotype and genotype and breaks down the activity as presented in Figure 14.6, with the aim of proposing a classification of possible errors. This is shown in Figure 14.7.

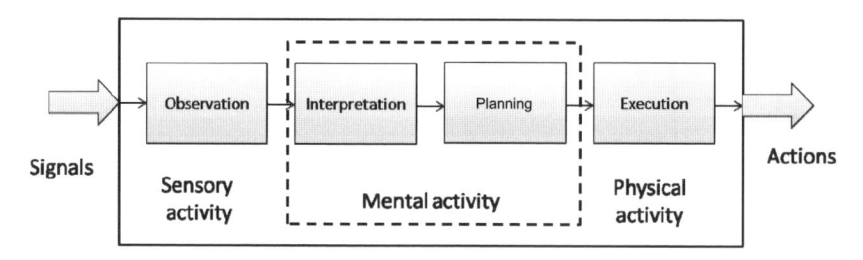

Figure 14.6. *Cognitive model used by the CREAM method*

14.4. Analysis and quantification of human errors

14.4.1. *Performance influencing factors*

The causes of human failure can be classified into different categories, as detailed in section 3.6. To consider these different aspects that influence human performance, most of these methods introduce the notion of "performance influencing factor".

Cognitive functions	Type of error	Explanation
Observation	Observation was not carried out	The observation of a signal or of an entity is not carried out (omission)
	Wrong observation	The wrong object is observed. The response is given to the wrong stimulus or event
	Wrong identification	A wrong identification is carried out due to an erroneous signal or to a partial identification
Interpretation	Faulty diagnosis	The diagnosis of the situation is incorrect or incomplete
	Reasoning error	A reasoning error (induction, deduction or choosing the wrong alternatives) leads to an invalid result
	Decision error	The decision is not made, a poor decision is made or a partial decision is made
	Delayed interpretation	The interpretation is carried out too slowly or is not fast enough given the conditions
	Poor forecast	The evolution has been incorrectly anticipated (unpredicted change, unanticipated side effects, poorly assessed evolution speed)
Planning	Inadequate plan	The plan of action is incompletely predicted (not enough details) or inadequate
	Priority error	A wrong target is selected in advance
Execution	Wrong time	The action is carried out too early, too late, or has the wrong duration
	Wrong execution	Poor execution related to force, distance, direction or speed
	Wrong object	Action carried out on the wrong object (neighboring object, similar object, or unrelated object)
	Wrong sequence	Action not executed (omission), or executed with missed stages or added stages, repeated stages or reversed stages

Figure 14.7. *Error classification based on cognitive functions*

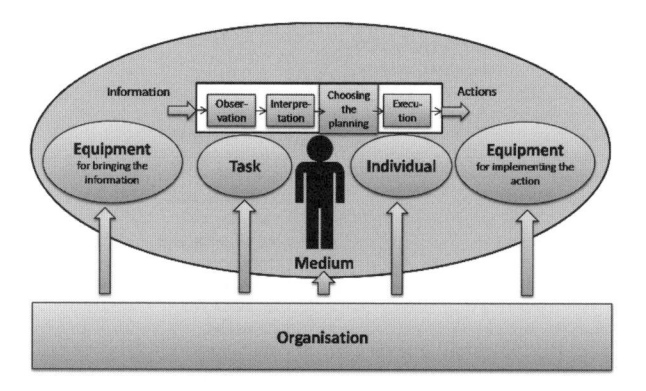

Figure 14.8. *Performance influencing factors (PIF)*

A PIF is defined as a factor that influences human performance and the calculation of the probability of human error. These factors can be external to the individual, or they may be a part of their characteristics.

The first list of these performance factors was given by Swain and Guttmann [SWA 83] within the THERP method (Figure 14.9). They are called *performance shaping factors* (PSF). We also find the denomination *error-producing conditions* (EPC) in the HEART method, K Factor in the TESEO method or *common performance conditions* (CPC) in the CREAM method.

External PSF's	External PSF's	Stress-related external PSF's	Internal PSF's
EXTERNAL: WORK SOLUTION	EXTERNAL: CHARACTERISTICS OF THE EQUIPMENT TASK	STRESS SOURCES: PSYCHOLOGICAL TENSION	INTERNAL: INDIVIDUAL FACTORS
Architectural features Quality of the environment (temperature, humidity, air quality, light, noise, vibration, general cleanliness) Work hours/work breaks Shift rotation Availability/suitability of equipment, tools and suppliers Manpower Organizational structure (authority, responsibility, communication channels) Manager's actions, work colleagues, workers' unions and policy specialists Bonuses, rewards and benefits	Perception-related requirements Physiological related requirements (speed, force, perception) Information display Anticipation requirements Interpretation Decision making Complexity (information load) Frequency and competitiveness Criticality of the task Long term and short term memorization Calculation requirements Information feedback Dynamic activities versus step-by-step activities Team structure and communication structure Factors in the human–interface (main equipment design, test equipment design or fabrication design, work aids)	Stress duration Task speed Task difficulty Important risk Fault risk, risk of job losses Monotonous work, degrading work, meaningless work Long, uneventful periods of vigilance Work performance related conflicts Absence of either encouragement or critique Sensorial deprivation Distractions (noise, glare, movement, flashes, colors) Illogical cueing	Previous training and experience State of current practice or skill Personality and intelligence Motivation and attitude Emotional state Stress state (mental or bodily tension) Knowledge of required performance levels Sex related differences Physical conditions Influence of family environment and of people or external organisms Group identification
EXTERNAL: WORK INSTRUCTIONS AND TASK-SPECIFIC INSTRUCTIONS		STRESS SOURCES: PHYSIOLOGICAL TENSION	
Required conditions (written or verbal) Warnings Work methods Site policy		Stress duration Fatigue Pains or discomfort Hunger or thirst Extreme temperatures Radiation Very strong acceleration Extreme pressure Lack of oxygen Vibrations Movement limitations Lack of physical exercise Disruption of circadian rhythm	

Figure 14.9. *Performance influencing factors (PIF)*

These factors allow us to describe the existence of potential causes of human error. The assessment of the importance of these factors, coupled with the relevant tables, allows us to assess or adjust the probability for the occurrence of human error, as we will see in the methods described in the rest of this chapter.

14.4.2. *Error probability assessment*

Although, throughout the years, numerous quantification techniques of human performance have been proposed there are no unanimously recognised methodologies based on a solid theoretical basis. Generally, the quantification can be carried out as follows:

– By breaking down the task into elementary tasks, as per the THERP method.

– By considering it globally, as per the HEART or CREAM methods.

The first type of method can seem more systematic, but it is not always simple to apply nor is it representative of human activity. What is more, it does not consider the cognitive aspect of the activity. The second type of approach is easier to implement, but it does not allow us to consider the specificities of the task and can cause inconsistencies between the assessments of different analysts.

Once the quantification procedure has been chosen, the determination of the probability can be carried out:

– by using data from a database such as Swain's handbook [SWA 83];

– by relying on expert judgment, which, as the name indicates, consists of appealing to the knowledge of specialists.

The inconvenience of the first approach is the necessity of detailed databases which, however, might not correspond to the context for a given application, whereas resorting to an expert judgment could itself lead to a large variation in the estimations.

Another aspect that must be considered for error quantifying is considering the time. This is the case, for example, with the *time reliability curve* (TRC) of Swain (Figure 14.14), or the HCR method. The main idea is that the more time a team of operators has, the lower the probability of an inadequate final reaction is.

Finally, for the majority of the methods, this quantification is done by including performance factors (PF), i.e. factors that influence human performance. This procedure is complicated by the fact that these factors are interdependent and these interdependencies are very difficult to estimate. Moreover, these factors are the result of a simplified vision of human behavior that applies to predefined classes of generic situations, and whose application to particular cases can prove problematic.

14.5. The SHERPA method

The SHERPA method was initially developed for the nuclear industry [EMB 86], but has since been used in several domains such as industry, aeronautics and even for mainstream products such as car radio systems. [STA 99]. This method allows for an identification of potential human errors and a qualitative assessment of their importance. The stages of this method are the following:

1) Building a model via HTA: this modeling allows us to describe the task that is being analyzed, by following the procedure described in section 7.7.1.

2) Classifying the tasks into one of the following categories:

    i) action;

    ii) information acquisition;

    iii) verification;

    iv) decision making;

    v) communication.

3) Identification of errors by using the taxonomy of the error modes proposed by the SHERPA method (Figure 14.10). For each task, the error modes given in the table are examined so as to identify the credible errors, which are explained as a phrase. For example, if the task is "checking that the reactor is empty", the error mode C1 will give the message "forgot to check that the reactor is empty".

Human Reliability Analysis 279

4) Consequence analysis: the objective of this stage is to determine and describe the consequences of the errors identified in the previous stage.

Type	Mode	Type	Mode
Action error	A1– Too long/too short operation A2– Poorly timed operation A3– Operation in the wrong direction A4– Too weak/too important operation A5– Misalign A6– Correct operation performed on the wrong object A7– Wrong operation performed on the right object A8– Omitted operation A9– Incomplete operation	Checking error	C1– Check omitted C2– Check incomplete C3– Right check on the wrong object C4– Wrong check on right C5– Check mistimed C6– Wrong check on wrong object
		Decision making error	S1– Selection omitted S2– Wrong selection made
Retrieval error	R1– Non-obtained Information R2– Wrong Information obtained R3– Information Retrieval Incomplete	Communication error	I1– Non-communicated Information I2– Wrong Information communicated I3– Incomplete Information communicated

Figure 14.10. *SHERPA taxonomy of error modes*

N°	Error mode	Description	Consequence	Recovery	P	C	Remedial strategy
1	C1	Forget to check if the reactor was empty	Explosion	None	L	H	Adding an alarm before loading if the level is not null

Figure 14.11. *Example of a SHERPA analysis table*

5) Analyzing the recovery possibilities: if it is possible to remedy the error in a following stage, this is indicated. If not, we indicate "none" in the table.

6) Assessment of the error probability: starting from the consequences and the recovery possibilities, the analyst assesses the error probability by using the following scale:

i) L:*low*, if the error has never taken place;

ii) M:*medium*, if it has already taken place but only rarely;

iii) H:*high*, if it takes place frequently.

7) Assessing the significance of the consequences of the error with the help of a three-level scale: *low*, *medium* and *high*, level H characterizing critical damage for the system.

8) Barrier analysis: this stage consists of analyzing the measures that must be implemented in order to reduce error possibilities. They can include actions at the level of the installation's design or the system's design. They are classified into four categories:

i) hardware equipment: modification, redesigning of the hardware;

ii) formation and training: implementing training exercises, new formations;

iii) procedures: review or implementation of new procedures;

iv) organization: modification in the culture or safety management policies.

The SHERPA method is a method that can demand a significant amount of effort for complex tasks and does not consider the cognitive aspect of human activity. However, it is a systematic and structured method for the analysis of human error. It depends on a taxonomy of generic error modes. It is relatively easy to apply and has been implemented successfully in various fields [BAB 96, SAL 03, STA 99].

14.6. The HEART method

The HEART method is a method that has been proposed by Williams in 1985 [WIL 85]. It is an approach that allows us to implement an assessment of the probability for human error in a way that is relatively simple and quick. Initially, it was mainly used in the field of nuclear industry, and then its use was extended to other fields.

The HEART method is based on a certain number of simplifying assumptions:

– The basic probability for human error depends on the generic nature of the task to perform.

– A list of nine categories (GTT) allows us to determine this probability in perfect conditions.

– If the conditions are not ideal, the probability for human error increases according to a coefficient determined by EPC factors characterizing the situation. A list of 38 EPCs is provided by the method.

The procedure is therefore as follows:

1) Building the model of the activity by using the HTA method, and in the SHERPA approach, by following the procedure given in section 7.7.1.

2) For each task of the model, determining its generic category (GTT) starting from a predefined typology (Figure 14.12), and, optionally, the confidence interval 5–95%.

3) Analyzing the situation in order to identify EPC factors (Figure 14.13), and their significance in the situation, in the form of a percentage. This work can be carried out in relation with the experts in the field.

4) Calculating the effect of each EPC:

$$\text{Effet}_{EPC} = (\text{Effect}_{EPCmax} - 1).\text{Proportion}_{EPC} + 1$$

The value of Effect_{EPCmax} is the one defined in the table, the proportion having been assessed in the previous stage.

EXAMPLE 14.1.– Let us consider a task of type D, that must be carried out in a limited timespan. The basic probability is $\text{HEP}_0 = 0.09$, the EPC factor for the limited time conditions is $f = 11$. The proportion of this EPC is estimated by experts at 40%. The final probability is then equal to $\text{HEP} = 0.09[(11 - 1)0.4 + 1 = 0.45$.

The HEART method has a certain number of limitations:

– The methodology used for identifying the EPC factors is not very well documented.

– It is based on the opinion of the analyst or experts on the choice of the type of task and the significance of the EPCs.

– The interdependence between the EPCs is not considered.

Generic task	Nominal HEP
A - Totally unfamiliar, performed at speed with no idea of consequences	0.55 (0.35-0.97)
B - Shift or restore system to new or original state on a single attempt without supervision or procedures	0.26 (0.14-0.42)
C - Complex task requiring high level of comprehension and skill	0.16 (0.12-0.28)
D - Fairly routine task performed rapidly or given scant attention	0.09 (0.06-0.13)
E - Routine highly-practiced, rapid task involving a low level of skill	0.02 (0.007-0.045)
F - Restore or shift a system to original or new state following procedures with some checking	0.003 (0.0008-0.007)
G - Completely familiar, designed, highly practiced task occurring several times per hour	0.0004 (0.00008-0.009)
H - Respond correctly to command even when there is an augmented or automated supervisory system	0.00002 (0.0-0.0009)
M - None of the above	Case by case basis

Figure 14.12. *Types of generic tasks in HEART*

However, its simplicity and ease of implementation have encouraged its use. An analysis carried out by Kirwan [KIR 97] has experimentally proven the validity of this method. It was used successfully for applications to various fields such as chemical industry, air transport, railway transport and medical applications [BEL 09].

14.7. The THERP method

The THERP method is one of the most widespread methods for assessing human reliability. It was developed by Swain in the 1960s [SWA 83]. The ASEP method is a simplified version of THERP [SAW 87].

EPC	Facteur
1- Unusual or unknown situation	17
2- Limited time	11
3- Poor signal-noise ratio	10
4- Means of deleting the information that is too easily accessible	9
5- No means of bringing the information to the operator in an accessible form	8
6- Gap between the mental representation model of the operator and that of the designer	8
7- No obvious means to reverse an unintended action	8
8- Overloading of transmission capabilities	6
9- Need to unlearn a technique in order to apply the opposite philosophy	6
10- Need to transfer knowledge from task to task without loss	5.5
11- Ambiguity over the requirements	5
12- Gap between perceived risks and real risks	4
13- Feedback of the ambiguous or wrong system	4
14- No clear confirmation of an action from the part of the system that the action is being carried unto	4
15- Operator's lack of experience	3
16- Poor quality of the information provided by the procedures and in the human-to-human interactions	3
17- Little or no independent oversight of outputs	3
18- Conflict between short term and long term objectives	2.5
19- Ambiguity over the required level of performance	2.5
20- Gap between the level of training required for the task and the operator's level of training	2
21- Incitement to use different, more dangerous approaches	2
22- Little occasions to use one's body and mind outside of work	1.8
23- Non-reliable tools	1.6
24- A need for judgments that are beyond the capacity or the experience of the operator	1.6
25- Unclear distribution of functions and responsabilities	1.6
26- No simple means of following the evolution of an activity throughout it	1.4
27- The possibility that limited physical capacities are surpassed	1.4
28- Little or no intrinsic meaning for a task	1.4
29- High emotional stress	1.3
30- Signs of poor health of the operators, particularly fever	1.2
31- Workmanship carried out with low morale	1.2
32- Incoherence in the meaning of displays and procedures	1.2
33- Bad and hostile environment	1.15
34- Prolonged inactivity or tasks of a low intellectual level being performed repetitively (first half an hour)	1.1
35- Next half an hour	1.05
36- Sleep cycle interruption	1.1
37- Work pace imposed by the intervention of others	1.06
38- Extra members of the team added on beyond the usual number of members that is needed to implement the task satisfactorily (per additional person)	1.03
39- Age of the staff in charge of perception tasks	1.02

Figure 14.13. Error Producing Conditions (HEART)

This method is well suited to cases where the tasks of the operators can be decomposed into elementary tasks, with an error that can be defined as a deviation from the specifications. It is less well adapted for analyzing complex situations involving tasks that must be carried out within limited time, such as diagnosing a problem during an accidental phase – where it is impossible to break down the task into elementary tasks.

This method is also better adapted to activities related to *skill-based* behavior, or *rule-based* behavior, than the activities of the cognitive type, i.e. the *knowledge-based* behavior. This method is based on usual methods for analyzing reliability of technical systems which are applied to human tasks.

First, we describe the sequence of the stages that the operator must carry out for the activity in question. Starting from this sequence, an event tree is built, each of these stages being associated with a pivotal event, with an alternative for the stage that was correctly implemented and an alternative corresponding to erroneous implementation (Figure 14.16).

For each elementary task, the guide of the method provides a value of the nominal probability for error, called *human error probability* (HEP). If the task is one for diagnosing the situation, a diagram provides a value of the HEP depending on time (Figure 14.14). For the other types of tasks, the omission and commission error probabilities are assessed on the basis of a set of tables, for each task category, given in Chapter 20 of the guide of the method [SWA 83]. An example is shown in Figure 14.15. A task that is carried out in an erroneous fashion can, in certain conditions, be corrected. It is the reason why the analysis procedure recommends the analysis of recovery possibilities, such as the verification by another operator, recovery during the following stage, the alarm on the erroneous action, or the periodic verification of actions. Specific tables provide the nominal probability for error of these recovery actions so they can be considered, either explicitly in the event tree, or in the calculation of the value of the final probability while using it in a multiplying way.

To consider the effect of the different parameters that influence human performance, Swain introduces the notion of PF, called the performance shaping factor (PSF). PSFs are classified into three main categories (Figure 14.9):

– External performance factor:
  - characteristics of the work situation;
  - characteristics of the job and task instructions;
  - characteristics of the equipment and the task.
– Internal factors that characterize the individual.
– Physiological and psychological stress factors.

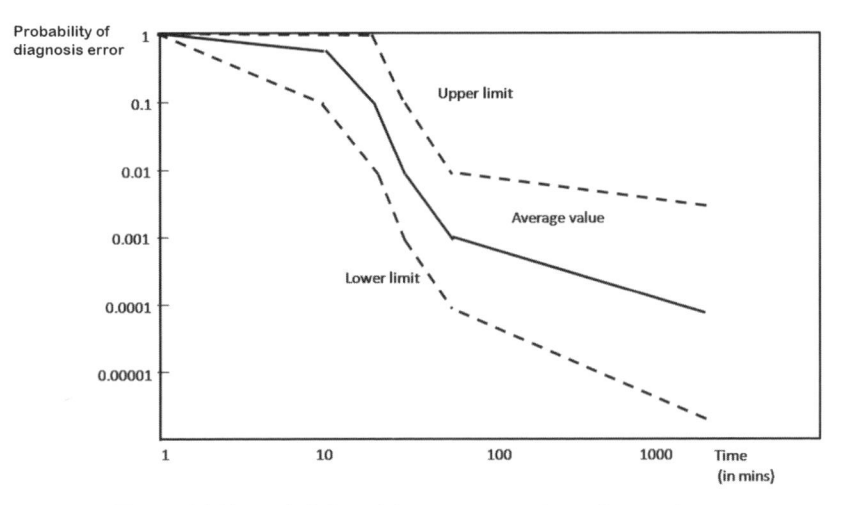

Figure 14.14. *Probability of diagnosis error depending on time (TRC: time reliability curve)*

	TABLE 20-13: NOMINAL ERROR PROBABILITY FOR THE SELECTION OF VALVES FOR LOCALLY OPERATED VALVES		
Item	Selection error in the case of a valve	HEP	Factor error
1	Clearly and unambiguously labeled, in a group of two or more valves, similar in size AND shape AND state AND marking	0.001	3
2	Clearly and unambiguously labeled, in a group of two or more valves, similar in size OR shape OR state OR marking	0.003	3
3	Not clearly and unambiguously labeled, outside of a group of valves similar in size AND shape AND state AND marking	0.005	3
4	Not clearly and unambiguously labeled, in a group of two or more valves, similar in size OR shape OR state OR marking	0.008	3
5	Not clearly and unambiguously labeled, in a group of two or more valves, similar in size AND shape AND state AND marking	0.01	3

Figure 14.15. *Example of probabilities table*

The tables provided in the guide of the THERP method provide the factors that help us modulate the value of the nominal probability for a certain PSF number, such as, the level of stress, whereas others must be assessed via expert judgment. The method also provides a procedure for optionally assessing the uncertainty on the estimated values.

The THERP method is applied to a scenario that was identified as potentially dangerous and for which it is necessary to quantify the probability of certain events that represent human errors. Once this is determined, the stages of the method follow:

1) building the activity model using the HTA method, as for the other approaches according to the procedure described in section 7.7.1;

2) identifying potential errors while using the omission/commission error taxonomy described in section 14.3.3;

3) building the event tree while considering these errors and finally other events of the system, which leads to a tree called *operator activity event tree* (OAET), which is described in section 7.7.3;

4) analyzing the HEP nominal probabilities for each task. Let us remember that the case of the task to establish a diagnosis is being specifically considered by using time reliability curve;

5) assessing the effect of the performance factors (PSF). Each PSF allows us to determine a corrective factor that will allow us to modify the value of the HEP in a multiplying fashion;

6) assessing the error probability of the set of the activities by exploiting the event tree.

EXAMPLE 14.2.– Let us once again take the example of Appendix 7. The operator's task in case of alarm consists of checking that the sprinkling system works, and, if it does not work, opening a valve manually. The operator is assigned to the monitoring task of the procedure, the work carries on in normal conditions, without any stress. The valve is not labeled, the opening/closing direction is indicated.

The task model of the operator in case of alarm is the following:

– diagnosing the situation;

– opening the sprinkling valve.

The first task is a diagnosis task, which needs a special assessment that considers time. The OAET model including the procedure fault is shown in Figure 14.16.

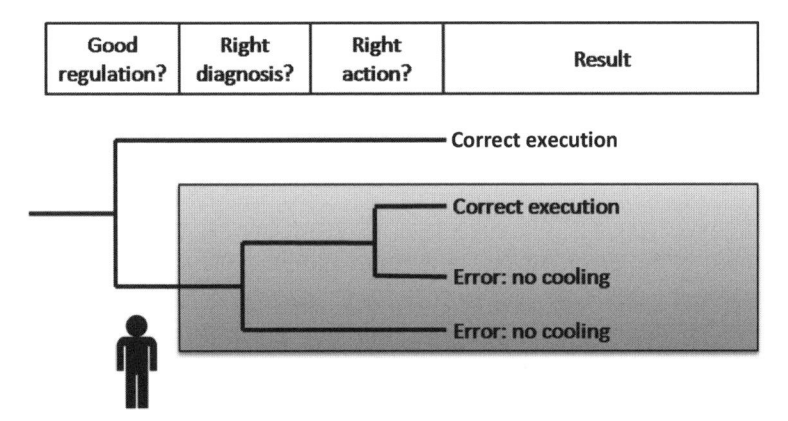

Figure 14.16. *Modeling of the operator's actions*

The simulation analyses have shown that the minimum time between the high-temperature alarm and an excessive pressure in the reactor that might trigger the rupture is 30 min. The time needed for the operator's actions is assessed by considering:

– the time needed for finding the procedure, should the operator have any doubts on what actions to take. This time is chosen to be equal to 5 min (Table 8.1 of the ASEP handbook [SAW 87]);

– the time for going to the valve and activating it, estimated at 3 min.

The time available for the diagnosis is therefore: $T = 30 - 8 = 22$ min. Using Table 8.2 of the ASEP textbook or Figure 14.14, the probability error for this task is estimated at 0.01.

Task 2 may be taken to be valve selection, and then its manipulation. The error probability is considered to be 0.005 (Figure 14.15).

The probability for human error is obtained by assessing the probability of a diagnosis error or a selection error after a correct diagnosis, and estimated at $P = 0.01 + 0.99 \times 0.005 = 0.015$.

This value can be used in a fault tree of the process.

14.8. The CREAM method

The CREAM method is a second generation method that seeks to better consider the cognitive aspects of the human activity. It was proposed by Hollnagel in 1998 [HOL 98]. It defines the notion of phenotype and genotype and relies on the cognitive model shown in Figure 14.6 for proposing a classification of possible errors.

Genotypes represent the categories of error causes, whereas phenotypes correspond to the error modes or error manifestations, that is, to what we observe. Each phenotype can be due to a cause that might belong to any genotype. In this type of classification, an error is therefore classified independently of its genotype or phenotype.

The genotypes belong to one of the following main categories:

G1) genotypes related to technological aspects: hardware, human–machine interfaces, procedures;

G2) genotypes related to the human aspect, particularly cognitive functions;

G3) genotypes related to organizational aspects, such as the organization, communication, formation, environment and working conditions.

The phenotypes are classified according to the manifestations along these parameters:

P1) date: too soon, too late, not at all (omission);

P2) duration: too long, too short;

P3) sequencing: stops before the end, does not stop at the end, bad command, omission, repetition, intrusion;

P4) object: neighboring object, similar object, bad object;

P5) force: too much, insufficient, irregular;

P6) direction: inverse, bad;

P7) amplitude/distance: too much, not enough;

P8) speed: too high, insufficient, irregular.

To facilitate the analysis, the phenotypes can be adapted to the four functions of the cognitive model of the operator (Figure 14.6), which leads to the modes listed in Figure 14.7.

The classification proposed by the CREAM method includes sub-categories for the phenotypes and for the genotypes that are defined in terms of general and specific "consequents" The classification also proposes connections between the consequents and the possible predecessors, which represent potential cause–effect connections and guide the analyst.

The CREAM method relies on a cognitive model (COCOM) that distinguishes between four modes of operator control. Following this mode, this will have little control on the situation or, on the contrary, will have the situation completely under control. The error probability of the task is then determined depending on the mode. The proposed modes and the probability intervals [FUJ 04] are the following:

– strategic: $0.00005 < p < 0.01$;

– tactical: $0.001 < p < 0.1$;

– opportunistic: $0.01 < p < 0.5$;

– scrambled: $0.1 < p < 1.0$.

To describe the context, the method introduces the notion of CPC (Figure 14.17).

The stages for the implementation of the basic version of CREAM are the following:

1) Building the model of the activity by using the HTA method, as for the other approaches, according to the procedure described in section 7.7.1.

2) Scenario specification: during this stage, the analyst specifies the scenarios that will make the object of the analysis. These can be connected to activities described in stage 1 or be operator responses to dangerous initiating events that can be represented as an event tree.

CPC	Importance/descriptors
1- Adequacy of the organisation	Quality of the definition of the roles and responsibilities, additional support, communication systems, safety management system, instructions, guides etc. **Very effective (+) /Effective/Ineffective (-) /Deficient (-)**
2- Working conditions	Nature of the conditions of physical work such as light, glare, levels of alarm, task interruption etc. **Advantageous(+)/Compatible/Incompatible (-)**
3- Suitability of the man – machine interaction and assistance	General man-machine interaction, including the information available on control panels, workstations, and the operational support provided by decision aiding mechanisms **Brings a plus (+)/Adequate/Tolerable/Inappropriate (-)**
4- Availability of procedures	The procedures and plans comprise operational procedures as well as emergency procedures, response heuristics with familiar schemes, routines etc. **Appropriate (+)/Acceptable/Inappropriate (-)**
5- Number of simultaneous goals	Number of tasks that a person must carry out at the same time (i.e. assessing the effect of the actions, sampling new information, assessing multiple objectives etc.) **Less than the possibilities/Adapted to the possibilities/More than the possibilities(-)**
6- Time available	Time available for carrying out the task, synchronizing the task with the dynamic of the procedure. **Adequate (+) /Inadequate from time to time/Always inadequate(-)**
7- Time of day (circadian rhythm)	Time of the day or night during which the task is carried out **Day/Night (-)**
8- Adequacy of training and experience	Level and quality of the training and experience of the operator **Adequate, high level of experience (+) /Adequate, limited experience/Inadequate (-)**
9- Quality of the collaboration between team-members	Quality of collaboration between the team-members including officials as well as non-officials, the level of confidence and the general social climate **Very effective (+)/Effective/Ineffective/Deficient (-)**

Figure 14.17. *CPC:* Common *performance conditions*

3) Context description: at this level, the analyst describes the context in which the activity or the scenario takes place. This description is built using the CPCs given in Figure 14.17. Each of them is rated subjectively, using the proposed descriptors. If, for example, the task regards the start up of a production following a clear and updated procedure, the CPC4 will be rated as "appropriate." The nine CPCs must be assessed.

4) Error prediction: for the scenarios identified, the analyst identifies the potential errors by using the phenotype/genotype classification or by using the most directly exploitable version adjusted for the cognitive functions given in Figure 14.7.

5) Quantification of the probability: if this is useful in the analysis, the error probability of the activity is assessed. This assessment is carried out via determining the control mode. This is obtained from the number of CPCs that have either a positive or a negative effect. The sign of the effect is obtained starting from the descriptor and the table. The number of positive effects and negative effects are used to determine the mode starting from the table

(Figure 14.18). For example, if we have three CPCs with (+) effects, and four with (-) effects, the mode obtained is "tactical", and the probability is in the interval $0.001 < p < 0.1$.

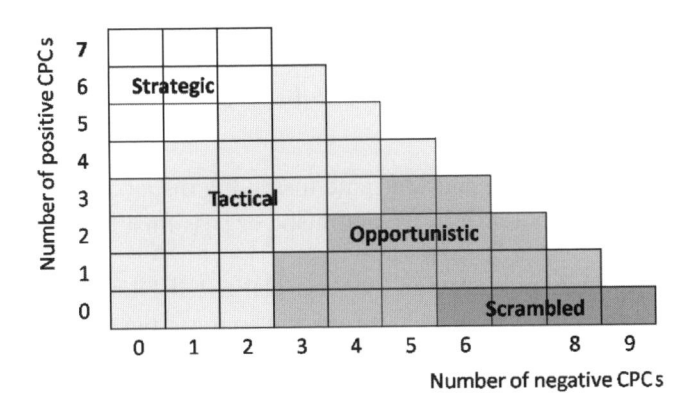

Figure 14.18. *Determination of mode in the CREAM method*

The extended version of the method uses a more detailed approach for assessing the probability [HOL 98].

The CREAM method is a method that can seem complex. One of its limits is that the validation is still currently underway [BEL 09]. It is, however, a second-generation method based on a cognitive model of the human operator, which is clear, systematic and well structured. Moreover, it can apply to different fields.

14.9. Assessing these methods

The assessment of human reliability can be carried out via a large number of methods. The results obtained with them may fluctuate significantly depending on the implementation conditions. From 1985 to 1987, the Joint Research Center (JRC) has carried out a comparative analysis between the different methods [POU 98]. This study considered two case tests:

– the routine maintenance and test task;

– the analysis of human activity during an accidental scenario that requires a diagnosing phase.

The THERP, SLIM, HCR and TESEO methods were part of the approaches that were being assessed. 15 teams from 11 different countries participated in the assessment. The results show:

– a dispersion of the results of more than 10^2 when the same method, THERP, was used;

– a dispersion of more than 10^6 between the results of the different methods;

– for the same team and the same method, a dispersion of 10^2 depending on whether the hypotheses were optimistic or pessimistic.

Another study published in 1997 on THERP, JEDI and HEART [KIR 97] has also shown discrepancies between the analysts but also a relatively high match between the estimated value and the expected value: in 30 analysts, 23 have obtained a probability value with a deviation factor below 10.

These methods must therefore be used with caution, but they also provide a systematic procedure for assessing the reliability of human actions.

Chapter 15

Barrier Analysis and Layer of Protection Analysis

15.1. Choice of barriers

The objective of risk analysis is to identify the events capable of creating damage and to determine the conditions under which these events might take place. Once these conditions have been identified, and if the risk level is deemed unacceptable, one of the risk-reducing approaches is the implementation of barriers. We will hereafter refer to these barriers as "safety barriers", and their aim is to:

– prevent the occurrence of damage (prevention measures);

– reduce the significance of damage (protection measures).

REMARK 15.1.– A certain number of the conditions leading up to the occurrence of damage are related to the barriers that might be failing.

The term "barrier" was introduced in the source–target model (Figure 15.1). A barrier allows us to stop the hazard flow, generated by the source, from reaching its target or to reduce the said flow. We can also represent the barriers in an event diagram, as we will show later on (Figure 15.8). Barriers limit the occurrence or modify the sequence of the generated events.

294   Risk Analysis

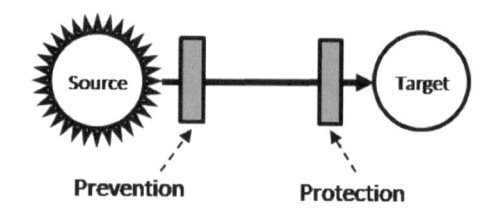

Figure 15.1. *Source–target model*

To carry out a comprehensive analysis, we use a bow-tie diagram that is the combination between a fault tree (Chapter 13) and an event tree. By using this formalism, the representation of barriers is made on the connections between the events. Some of the effects of a barrier are:

– reducing the occurrence probability of the event that follows it;

– reducing the severity of an event, that is making the events that follow the barrier become altered and less severe.

In a bow-tie diagram, the first interpretation (probability reduction) is generally the one that is used. We note that the event following the barrier is the one that takes place when the barrier has a failure (Figure 15.2). It is the event that would have taken place had there not been any barrier. Should there be a barrier, the probability of the event is reduced, because the event only takes place when the barrier has a failure. If the barrier functions properly, there is still a possibility for damaging consequences, although usually they are less significant. These consequences are represented by a connection that starts from the bottom the barrier. The new event symbolizes the less severe damage.

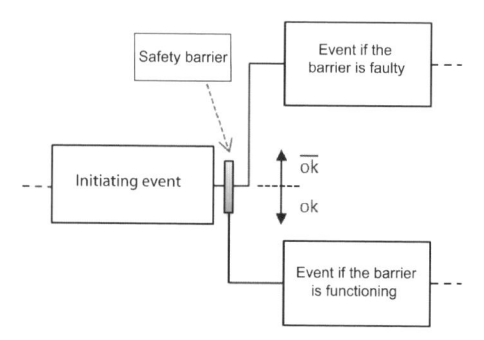

Figure 15.2. *Modeling a barrier via an event graph*

15.2. Barrier classification

Several approaches were proposed for barrier analysis [HAD 73, HOL 04, SAL 06, SKL 06a]. First, it is worth asking in what way the barrier acts in relation to the accidental sequence. In this sense, a classification was proposed within the ARAMIS method [SAL 06]. We distinguish between the barriers that allow us to carry out the following:

– A risk suppression operation: a modification in the design causes this type of scenario to disappear.

– Risk-prevention: the barrier reduces the occurrence probability of the initiating event.

– A control of the evolution: the barrier limits the deviations so as not to lose control.

– Target protection: the barrier limits the effect of the damages by protecting the target from dangerous phenomena.

These types of barriers are found in different stages of the accidental sequence, as illustrated in Figure 15.3.

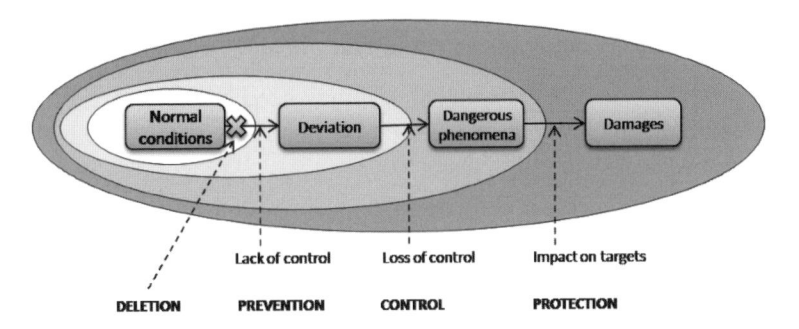

Figure 15.3. *Barrier classification following the stage of the accidental sequence*

The first type of barriers is not actually implemented in the system for which we are trying to reduce risk, but has led to the modification of the system. For the others, the way of actually implementing the barrier can be classified according to the diagram in Figure 15.4. This classification is based on the following properties:

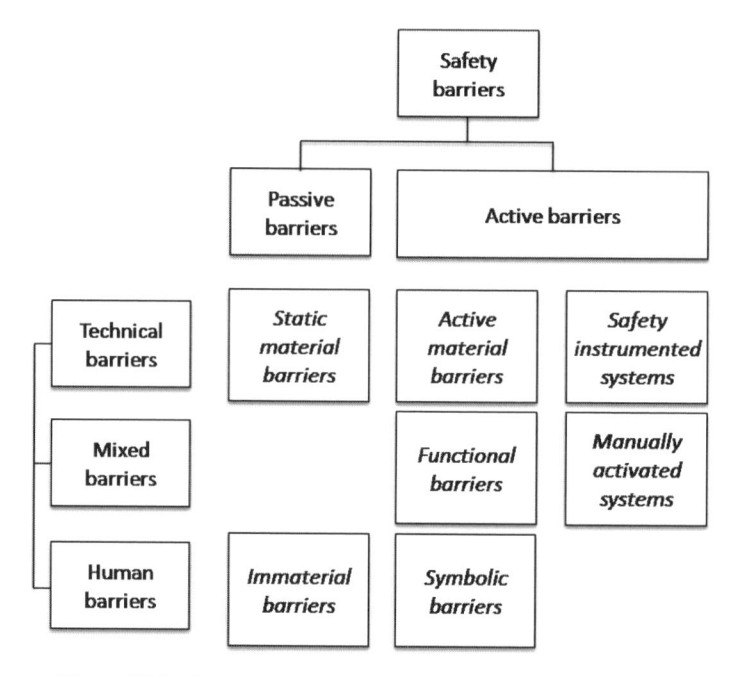

Figure 15.4. *Classification of the barriers according to their types*

– A barrier may be active or passive. A barrier is said to be passive when it does not put into play any mechanism in order to act.

– A barrier can be purely technical or it may need human intervention. It is then called "human". Some of them are mixed barriers, such as manual safety systems.

– A barrier may or may not be material [HOL 04].

Starting from these properties, we can distinguish between the following:

– Static material barriers: these are the barriers that limit, through their presence, the energy flow or the material flow between a source and a target, for example containments, cages, sumps and dikes.

– Other material devices: these barriers are active, but in a mechanical way, such as safety valves or rupture disc reliefs.

– Safety instrumented systems (SISs): these barriers are automatic systems for shutting down the installation (in safe mode) upon detecting an undesired event.

– Functional barriers: these barriers limit the possible functions of a system at a given point by using several blocking mechanisms, for example locks, single-key systems, passwords and codes.

– Manual action systems, which are mixed barriers with both technical and human components: the operator interacts with the technical elements of the safety system that it oversees or upon which it acts, for example a valve that needs to be closed at a high temperature.

– Immaterial human barriers: these barriers are not physically present, but rather are principles or knowledge such as the competence of the operator, regulations and safety principles.

– Symbolic barriers: these barriers need a human interpretation, for example traffic signs, floor marks, instructions and working permits.

SISs are very important systems in terms of barriers. Part of this chapter is dedicated to SISs (section 15.5). We can also see that this classification does not explicitly highlight the organizational aspect. In fact, this only manifests itself at the end, either via risk suppression or via a technical or human barrier.

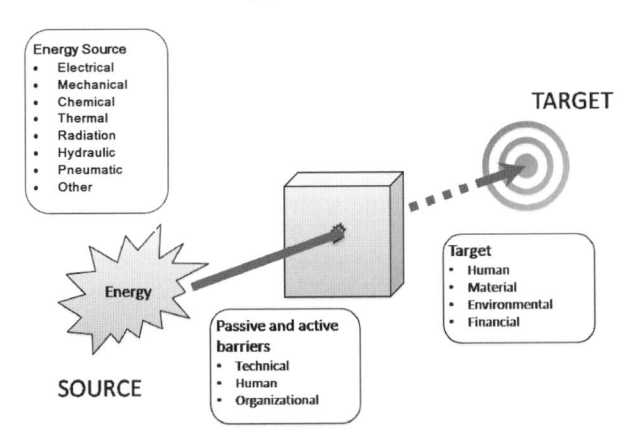

Figure 15.5. *Implementing barriers via the source–target model*

15.3. Barrier analysis based on energy flows

This approach is based on Haddon's work carried out in the 1970s [HAD 73]. The basic model used is the one shown in Figure 15.1. The main idea is to implement barriers that will limit or stop the energy sources, which are sources of hazard, generating damage on the targets. The stages of this analysis are as follows:

1) Identifying the danger sources, which can be carried out either via the preliminary hazard analysis (PHA) or the systemic and organized risk analysis method (SORAM) method.

2) Identifying the barriers suited for each type of source.

3) Assessing the ability of each of these barriers to fulfill their role, i.e. to stop the hazard from reaching its target.

The barriers are classified depending on their function, their location and their type (Figure 15.6). To guide the barrier analysis, the method suggests the following principles:

– Avoid the concentration of energy or limit its level and quantity. For example, by using smaller containers.

– Stop the release. For example, by using jacketed containers.

– Modify the rate of release of energy. For example, by placing rupture disc reliefs.

– Separate the energy source and the target in terms of time or space. For example, by evacuating the dangerous areas or controlling access to the dangerous area.

– Isolate the source by installing a physical barrier. For example, with the help of cages and containments.

– Protect the target. For example, by using an isolating system or personal protection equipment.

The assessment of damage probability and its severity is then performed with the proposed barriers so as to validate them. In the case where the risk level remains excessive, new barriers must still be added. Moreover, it is worth checking that the barriers do not create new risks.

The main limit of this approach is that it only considers the aspects related to random releases of energy potentials. Certain cases are not easy to formulate this way, such as anoxia or damage generated by utility loss. Moreover, this type of analysis does not allow us to easily consider earlier causes that lead the system into a dangerous situation. It is, however, well suited for simple cases.

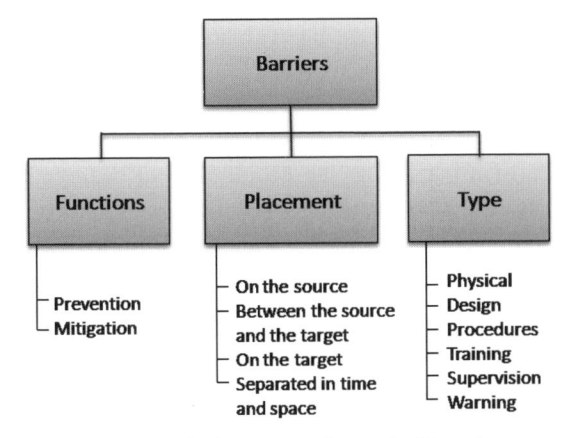

Barrier Analysis and Layer of Protection Analysis 299

Figure 15.6. *Classifying the barriers in the method based on energy flows*

15.4. Barrier assessment

The assessment of a barrier differs a little depending on whether the device is active or passive. The assessment stages are the following:

1) Adequacy and independence: the barrier must fill the expected role, i.e. must minimize the consequences of the specified initiating event and not be affected by the accidental phase. For example, if a safety loop seeking to limit the temperature is connected to the same sensor as the one used for regulation, this safety chain cannot be considered to act as a safety barrier (detection part) for a critical event initiated by a failure in temperature regulation, because it will fail if the sensor breaks down.

2) The robustness for the use in safety: the barrier must have a simple design and be robustly implemented.

3) The following stage consists of verifying its efficacy in a usage context and for a given period of functioning. A barrier can be 100% efficient or not efficient at all. In this case, it is worth describing the situation obtained with the functioning of the barrier. For example, a barrier can only function for a maximum period of time, like a fire door does, or only stop part of the phenomenon. A water screen can only stop a certain fraction of a cloud. In this case, in the bow-tie diagram, we will have to represent the product-releasing event with a reduced flow.

4) If the device is active, we must check that the response time is appropriate.

5) The last parameter that must be evaluated is the confidence level of the barrier or the failure probability when it is needed – an aspect that we will detail later on.

The influence of a barrier on the probability of the event that follows it can be modeled by an AND gate between the previous event and an event representing the failure of the barrier (Figure 15.8). In this figure, the C2 event is the one that takes place should the barrier function well and C1 is the event that takes place in the case of the barrier's failure. To evaluate the influence of a barrier on the event that follows it, it is therefore necessary to know the probability of failure of the barrier. We distinguish between two cases: the systems operating in low-demand mode and those operating in high-demand mode. In the first case, we use the notion of *probability of failure on demand average* (PFD_{avg}), whereas in the second case, we use the failure rate of the device in order to evaluate its failure probability after a given period of time (Appendix 4); this failure rate is assumed to be constant.

A similar index to the safety integrity level (SIL) defined in the rest of this chapter was proposed by INERIS [INE 08] in order to characterize the level of confidence. It is an integer, denoted by NC, which varies from 0 to 4. It is related to the PFD by the relation:

$$10^{-(NC+1)} \leq \text{PFD} < 10^{-NC}$$

The values are explained in Figure 15.7. This figure is based on the NF standard EN 61511-1 for the functioning mode on demand (where the SIL level was replaced by NC and the NC0 line was added).

The NC indicator allows us to characterize qualitatively the confidence level of a barrier. For example, an $NC = 2$ barrier will have a PFD_{avg} between 10^{-2} and 10^{-3}. Besides its PFD, other requirements must be fulfilled for a piece of equipment to get the qualification at a given SIL level, as will be detailed in the following. The PFD is used to assess the risk reduction factor, denoted by RR. We give an example at the end of this chapter using the layer of protection analysis (LOPA) method.

Barrier Analysis and Layer of Protection Analysis 301

CL Confidence level	Average probability of failure on demand (PFD$_{avg}$)	Risk reduction (RR)
4	$10^{-5} \leq PFD_{avg} < 10^{-4}$	$10\,000 < RR \leq 100\,000$
3	$10^{-4} \leq PFD_{avg} < 10^{-3}$	$1\,000 < RR \leq 10\,000$
2	$10^{-3} \leq PFD_{avg} < 10^{-2}$	$100 < RR \leq 1\,000$
1	$10^{-2} \leq PFD_{avg} < 10^{-1}$	$10 < RR \leq 100$
0	$10^{-1} \leq PFD_{avg} < 1$	$1 < RR \leq 10$

Figure 15.7. *Correspondence between the confidence level and the risk reduction for low demand systems (INERIS source)*

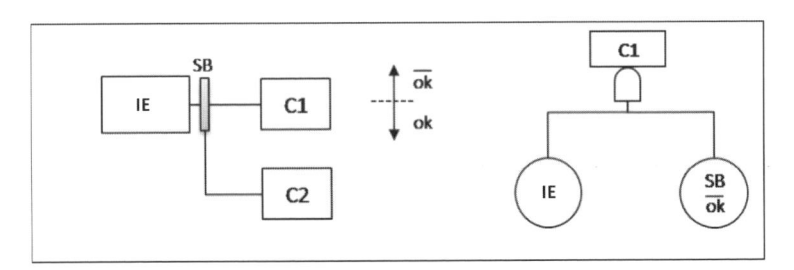

Figure 15.8. *Modeling of a barrier via an event graph*

15.5. Safety instrumented systems

15.5.1. *Introduction*

Of all safety barriers, there is one category that is particularly important: they are called safety instrumented systems (SISs). These are automatic systems designed to react to dangerous events. They are made up of (Figure 15.9):

– one or several sensors;

– one or several processing units (safety automata, embedded systems);

– one or several actuators (valve actuators or other actuators).

An SIS is used for implementing one or several *safety instrumented functions* (SIF). It is a function whose objective is to maintain or control the equipment (*equipment under control* (EUC)) in safety upon detection of an event or a predefined deviation. For each safety function, the components of an SIS allow it to implement the following three sub-functions:

– Detection, performed by a sensor, which is the element that allows the transformation of a physical information (pressure, temperature, flow, concentration, etc.) into an electrical measurement suited to the processing. This measurement is conditioned into a signal by a transmitter in order to be transmitted to the processing unit. The signal transmitted can be an analog 4–20 mA signal or a binary signal of the "all or nothing" type (AON, or 1/0).

– The processing, more or less complex. It can be limited to a simple display or a comparison for generating an actuator's command. We may distinguish between two main technologies, which differ in their level of reliability and their treatment ability:

- the wired technologies, based on elementary logic components, such as relays, connected with each other either electrically, pneumatically or hydraulically;

- software based technologies, based on data acquisition units or alarms, programmable logic controllers (PLC), some of which can be dedicated to safety (Safety PLC), or even industrial computers or electronic cards with microprocessors or with logic solvers.

– The action, which is carried out by actuators that transform a signal (electrical, pneumatic or hydraulic) into a physical phenomenon that allows us to command an organ such as a pump and a valve. Depending on the type of mechanical energy, we talk of an electric actuator, a pneumatic actuator or a hydraulic actuator. The actuators are coupled with terminal elements: valves, pumps, alarms, etc.

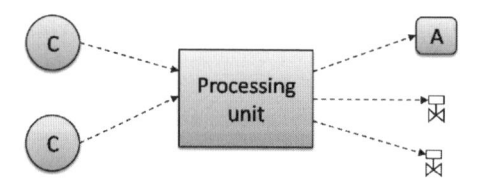

Figure 15.9. *Example of an SIS architecture*

These elements communicate with each other via various technologies: electric cables, electromagnetic waves, optic fiber, pneumatic or hydraulic pipes.

The SISs are more and more numerous in the systems that surround us: emergency stopping of chemical procedures, detection and automatic sprinkling in case of fire, gas detection, automatic train stopping, air bags, on-board pilot assistance (ABS), safety systems for medical purpose, radiotherapy devices, etc.

15.5.2. *IEC 61508 standard*

A certain number of standards have been established for the design of the SISs. The general standard is the IEC 61508 standard published in 2002. It is entitled "Functional safety of electrical, electronic, and safety-related programmable electronic systems", and it is made up of seven parts that describe how to manage functional safety for equipment when using an SIS. These parts are entitled:

– part 1: general requirements (lifecycle);

– part 2: requirements for the E/E/PES systems related to safety;

– part 3: software requirements;

– part 4: definitions and abbreviations;

– part 5: example of methods for determining the levels of safety integrity;

– part 6: guidelines for implementing parts 2 and 3;

– part 7: overview of techniques and measures.

This standard has been adjusted to different fields:

– Process industry: IEC 61511 or ANSI/ISA-84.00.01-2004, functional safety, SISs for the field of production via process.

– Nuclear: IEC 61513, nuclear plants – instrumentation and command control of the systems that are important to safety.

– Machines: IEC 62061, functional safety of electric command systems, electronic and safety-related programmable electronics.

– Railway: IEC 62278, railway applications – specification and reliability proof of the availability, maintenance and safety (RAMS).

– Automobile: ISO 26262, road vehicles – functional safety.

– Electro-medical equipment: IEC 60601, general requirement for the basic safety and essential performances.

Standard 61508 is a standard that providers rely on. It is particularly possible to acquire a system guaranteeing a given SIL level. We explain, in what follows, the principle that this standard proposes for determining the integrity level of an SIS.

15.5.3. *Failures of an SIS*

The more a scenario can generate damaging consequences, of significant severity, the more the safety system allowing us to reduce the consequence probability must be reliable. The SIL is a performance measure of the reliability level of this safety system.

DEFINITION 15.1.– *The SIL is defined by an index characterizing the probability of good functioning of a safety device, under all the stated conditions, for a given period of time (standard 61508). The relations between levels and probability are defined in Figure 15.10.*

SIL	SIS Demand		Risk reduction factor
	low PFD_{avg}	high PFH	
4	$[10^{-5}, 10^{-4}]$	$[10^{-6}, 10^{-5}]$	10 000 to 100 000
3	$[10^{-4}, 10^{-3}]$	$[10^{-6}, 10^{-5}]$	10 000 to 1000
2	$[10^{-3}, 10^{-2}]$	$[10^{-6}, 10^{-5}]$	1000 to 100
1	$[10^{-2}, 10^{-1}]$	$[10^{-6}, 10^{-5}]$	10 to 100

Figure 15.10. *The SIL levels*

For a given scenario, in order to obtain a given risk level and considering the probability of the initiating event and the severity of its consequences, standard 61508 defines the risk reduction rate that is necessary, such as the SIL that the safety system must achieve. Several approaches are proposed for determining this level:

– An approach based on a decision graph (Figure 15.11) that allows us to choose the SIL level in a simplified manner depending on the significance of the consequences, the occurrence frequency of these consequences when there are no barriers, the possibility of avoiding consequences and the desired final level of probability, denoted by W.

– An approach based on the LOPA method (see section 15.6).

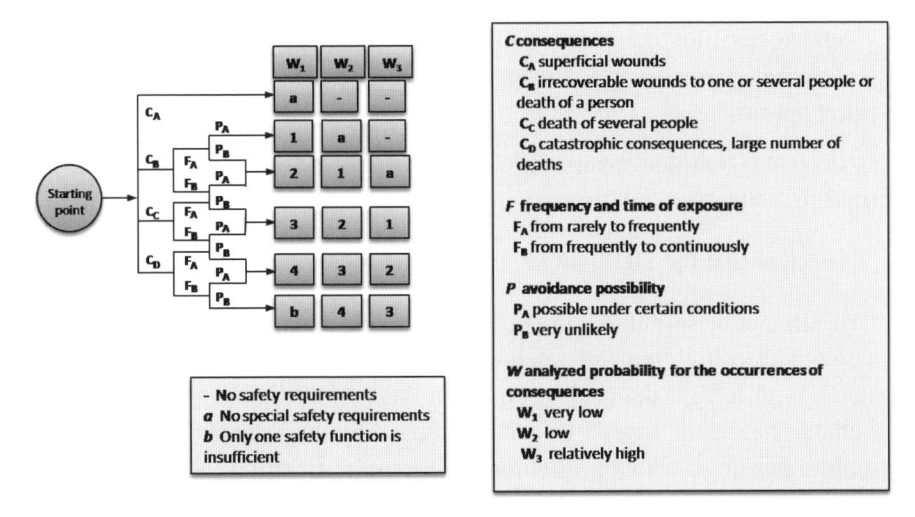

Barrier Analysis and Layer of Protection Analysis   305

Figure 15.11. *Decision tree for the choice of the SIL level*

EXAMPLE 15.1.– Let us suppose that the SIS is used as a barrier for a phenomenon whose consequences can go as far as the death of several people, that the system is used often and that the possibilities for avoiding damages are quite unlikely. These conditions correspond, for example, to an electronic braking system in a car. The level for C is C_c, the level for F is F_B, the level for P is P_B. If we wish to have a very low damage probability, say W_1, the required SIL level according to the graph is 4.

The failure analysis of the SIS allows us to check whether the system achieves the required level or not. In this approach, we distinguish between two main failure modes for a safety function (SIF):

– A failure mode on demand: the function is not carried out when necessary.

– A failure mode connected to random functioning: the function is triggered without the presence of the event to which it is supposed to react.

The failures of the different elements of the SIS can be classified into four categories:

– undetected dangerous failure (*dangerous undetected* (DU)) that we only detect upon demand or during a test run; examples of DUs include a fire sprinkling valve, which is blocked shut;

– detected dangerous failure (*dangerous detected* (DD)), which is detected via an auto-test;

– undetected non-dangerous failure (*safe undetected* (SU)), which is not detected, but it is also not dangerous, such as the blocked brakes during a parking period;

– detected non-dangerous failure (*safe detected* (SD)), such as a fire sprinkling valve blocked open.

It is clear that the DU-type failures are the most problematic.

An SIS can be solicited more or less often. We can distinguish between the systems with a low demand, such as emergency stop or an air bag, and the systems with a high demand, such as an ABS system on a car. The former are characterized by their PFD, and the second by their dangerous probability of failure per hour, called a PFH index. A usual estimation, because the PFH value is low, consists of considering that the annual failure probability is given by $P_{def} = \text{PFH}.10^4$, the number of hours in a year being close to 10^4.

The standard defines four SIL levels depending on the PFD$_{avg}$ or on the dangerous probability per hour PFH. These sizes are associated to a risk reduction factor (Figure 15.10). Let us mention that this factor is only an element for defining the SIL level of an equipment. To enter this category, standard 61508 defines other criteria that the equipment must verify, particularly qualitative criteria such as the architecture or the implementation [IEC 10a].

The PFD$_{avg}$, for low demand equipment, can be determined as a function of the failure rate in an undetected dangerous mode λ_{DU} and in the duration between two tests, denoted by T_{test}. If the element is started at the time $t = 0$, the PFD at the time t is given by (Appendix 4):

$$\text{PFD}(t) = 1 - e^{-\lambda_{DU} \times t}$$

The calculation of the average value of PFD(t), between two testing periods, and the moment at which it goes back to zero, is written thus:

$$\text{PFD}_{avg} = \frac{1}{T_{test}} \int_0^{T_{test}} (1 - e^{-\lambda_{DU} \times t}) dt$$

$$= 1 - \frac{1}{\lambda_{DU} T_{test}} (1 - e^{-\lambda_{DU} \times T_{test}})$$

Because $\lambda_{DU}T$ is a small number, we use the following approximation:

$$\text{PFD}_{avg} \approx \frac{\lambda_{DU}T_{test}}{2}$$

This relation is true for a non-redundant, periodically tested SIS. It is often used as a basic value in the bow-tie diagram analyses or in fault trees. When redundancies exist, or if we consider common cause failures, often with the approach β-factor, we use other relations [GOB 10, GRU 05].

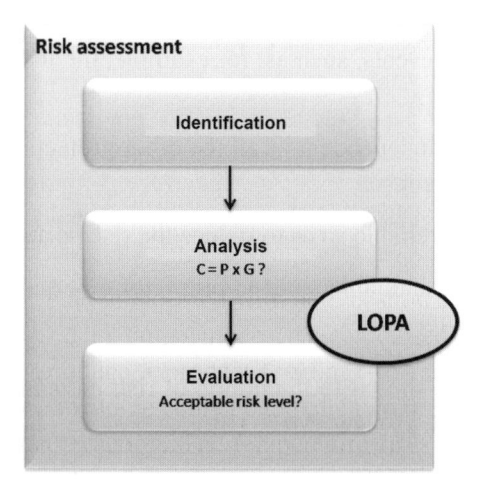

Figure 15.12. *The LOPA method in the risk assessment process*

15.6. The LOPA method

15.6.1. *Description*

The LOPA method was developed at the end of the 1990s by the *Center of Chemical Process Safety* (CCPS). The objective of this method is to assess the risk reduction level with existing barriers on a system and to decide whether additional barriers must be added. It is defined by its authors [AIC 01] as a semi-quantitative simplified method, because it is based on levels defined by multiples of 10 to represent the probabilities, the frequencies and the severity levels. This approach is similar to the one used in the French regulation regarding industrial risks.

This method is an analysis tool that is used at the end of a qualitative risk identification method, such as a preliminary risk analysis, a PHA or a hazard and operability (HAZOP) analysis. A key concept in the LOPA approach is the notion of an independent protection layer (IPL). The method seeks to determine if sufficient protection levels are provided to ensure that the risk level is tolerable.

An IPL is defined as a system implementing a safety function, either actively or passively, and that must verify the following criteria:

– An IPL must allow us to detect and prevent or mitigate the consequences of specific dangerous events, such as, for example, the loss of containment or a reaction explosion.

– An IPL must be independent of all the other protection layers associated with the dangerous event.

– An IPL must be reliable so as to reduce the risk of a specified level.

– An IPL must be auditable so as to allow for a periodic validation of the safety functions that it ensures.

The IPL notion is similar to the notion of risk reducing measures (RRM) defined in the French 2010 law [MEE 10].

Let us cite some IPL examples:

– standard operational procedures;

– process control system;

– alarms with an acknowledgment of the operator;

– SISs;

– safety valves;

– anti-explosion walls;

– fire and gas sensors;

– sprinklers.

The LOPA method is a method capable of ensuring that the risk level is kept under control at an acceptable level. This is a rational methodology, which is based on a well-argued process, that allows for a quick identification of the protection levels, which diminishes the frequency and/or the consequence of the initiating dangerous events. It proposes an approach for the assessment of the barriers in order to eliminate the subjective aspect of the qualitative methods with a cost-efficiency ratio better than that of a purely quantitative method.

The LOPA method supposes that a preliminary identification of the scenarios leading up to damage was already carried out. Each of the scenarios analyzed is part of a dangerous initiating event leading to a consequence. For this scenario, we determine what the existing independent safety barriers (IPL) are, whether they are of a technical, organizational or human nature. Figure 15.13 presents these barriers as successive layers.

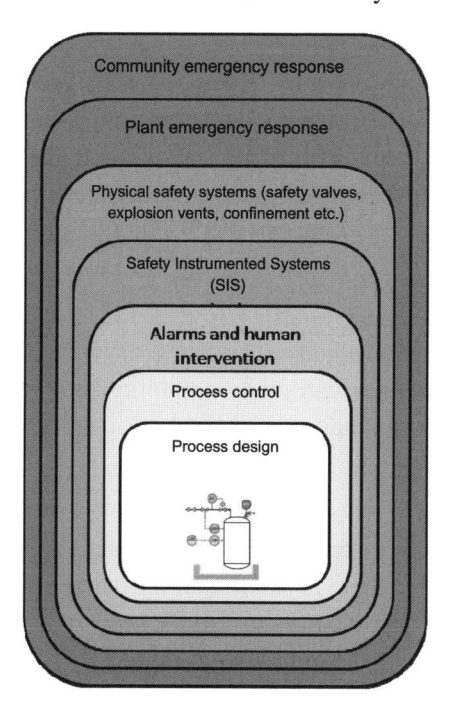

Figure 15.13. *Independent safety barriers (IPL)*

The barriers are represented in the scenario. The probability of the scenario is assessed via an event tree, allowing us to estimate the different possible consequences depending on the correct functioning of the barriers. The tree is reduced, because it suffices that early barrier functions for undesirable consequences are to be avoided, as it is represented in Figure 15.14.

310   Risk Analysis

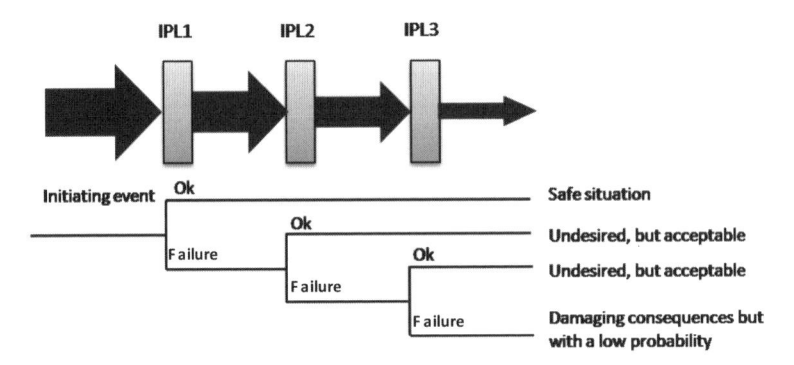

Figure 15.14. *Comparison of LOPA and event tree analysis*

The stages of the method are the following:

1) Establishing the selection criteria of the scenarios: this stage is implemented in the task "establishing the context", prior to the risk assessment stage. It is important because it allows us to limit the time needed for the analysis. One rule could be to only consider the scenarios that are significant in terms of consequences either by relying uniquely on the intensity of the phenomenon or by combining this intensity with its vulnerability.

2) Scenario development: starting from a risk analysis such as PHA, the scenarios to be built are selected according to the criteria defined above. They are completed by using a hazard and operability study (HAZOP) analysis or failure mode and effects analysis (FMEA) to identify the earliest initiating events.

3) Identification of the frequencies of initiating events: the frequencies of initiating events are estimated on the basis of internal experience feedback and the information available in the reliability databases or in the documents published by inter-professional organizations such as [ICS 09] (Appendix 5).

4) Identification of safety barriers (IPL) and of their PFD: for each scenario, a detailed analysis allows us to identify the existing safety barriers by checking that these devices are independent of the others and also independent of different failures and dangerous phenomena that can take place within the system (especially common cause failures). The PFD is assessed starting from predefined rules (see section 15.6.5). The LOPA method relies on SILs in order to characterize the IPLs.

Barrier Analysis and Layer of Protection Analysis 311

5) Risk level calculation: the probability of the final event of the scenario, the one that characterizes the consequences, can be calculated from the probabilities of the initiating events and from the PFDs of the IPLs. It is therefore possible to determine the level of risk by using a risk matrix or any other means chosen by the analyst.

6) Risk assessment and choice of treatment that must be carried out: this stage consists of comparing the risk level of the scenario to the one settled in part 1 for determining whether this risk is tolerable or not. If necessary, new IPLs will be added. In the decision-making process, we will prefer several IPLs with a low SIL level that one IPL with a high SIL level, the first solution being more tolerant to uncertainty.

The main limits of the method are related to the relative difficulty for analyzing a scenario compared to a qualitative method. The choice of scenarios must therefore be made carefully. Moreover, it is important to check that the different barriers are truly independent, because this hypothesis lies at the core of the method.

15.6.2. *Scenario identification*

This work is carried out after a risk analysis of a PHA, HAZOP or FMECA type. The dangerous events (or their causes) that could potentially lead to severe consequences are selected as initiating events. The accidental scenario can be represented as an event tree (Figure 15.14). In this illustrative figure, the thickness of the arrow represents the frequency of the occurrence.

Because the protection levels are supposed to be independent, the frequency of the final event can be obtained by following the same approach as the event trees or the bow-tie diagram:

$$f_C = f_i \times \text{PFD}_1 \times \ldots \times \text{PFD}_n$$

where f_C is the frequency of the consequences in case of failure of all the barriers (IPL), f_i is the frequency of the initiating event and PFD_i is the PFD of the barrier i.

The set of scenarios is given in a table as presented in Figure 15.15, made up of the following columns:

1) Description of the consequences of the scenario, in the case where no barrier functions.

2) Severity of the scenario.

3) The initiating event of the scenario.

4) Its frequency, which is generally given per year, that is used in the LOPA method to quantify the likelihood of the occurrence of the initiating event.

5) Barriers related to the process design.

6) Barriers implemented by the BPCS (Basic Process Control System).

7) Barriers implemented as alarms and procedures that must be followed by the operators.

8) SISs, automated systems dedicated to safety.

9) Others barriers, such as safety valves, limited access, cages, containers.

10) The number of independent barriers cited in the previous columns.

11) The frequency of the consequences described in column 1 and with the severity noted in column 2. This is obtained by multiplying the frequency of the initiating event by the PFDs of the different independent barriers.

N°	Consequences		Initiating event		IPL					Consequences
	Description	Level of severity	Initiating event	Frequency (/year)	Process design	BPCS	Procedures, alarms	SIS	Other barriers	Number	Frequency (/year)
	(1)	(2)	(3)	(4)	(5)	(6)	(7)	(8)	(9)	(10)	(11)

Figure 15.15. *LOPA scenarios table*

15.6.3. *Analysis of the scenarios*

The assessment of the severity is implemented by an approach similar to other methods, following an approach as the one described in section 6.6.2. It is possible to use:

– an assessment of the intensity of the phenomenon, such as the significance of the leakage or of the fire, without assessing the actual impact on the population;

– a qualitative assessment of the consequences, by considering a table like the one presented in Figure 15.16 [FRE 06];

– a quantitative assessment of the consequences, by assessing, for example, the number of people that have actually been affected, taking into account a map of stakes and using a model of the effects for determining the effect zones, following the approach described in section 6.6.2.

Severity of final consequences	Consequences
Minor	Impact limited to the area local to the phenomenon with larger potential consequences if no action is taken
Serious	Impact capable of causing serious injury or death on site or outside of the site
Major	An impact five times more significant than the 'serious' level

Figure 15.16. *Severity levels*

The LOPA method does not suggest one approach in favor of another [AIC 01]. The choice depends solely on the context of the analysis.

15.6.4. *Identification of the frequency of initiating events*

The initiating events can be of different types. We can classify them into:

– failures related to equipment and products, such as mechanical faults, wear, vibration problems, clogging, misuse, a deviation of the properties of the product and software bugs;

314 Risk Analysis

– human errors, such as operation errors, driving errors, maintenance errors, the errors in response to alarms;

– external events, natural or man made, such as floods, earthquakes, plane crashes, accidents on neighboring installations.

A more detailed presentation of these categories is given in Chapter 3. The frequencies of these initiating events are determined in using the reliability databases (Appendix 5) or specific tables for the LOPA method [AIC 01]; we offer an example below (Figure 15.17).

Initiating event	Annual frequency (literature) (per year)	Value chosen for LOPA (per year)
Pressure vessel failure	10^{-5} to 10^{-7}	10^{-6}
Piping failure (full breach)/100m	10^{-5} to 10^{-6}	10^{-5}
Pump seal failure	10^{-1} to 10^{-2}	10^{-1}
Pressure vessel failure	1 to 10^{-1}	10^{-1}
Operator failure (to execute routine procedure)	10^{-1} to 10^{-3} (per operation)	10^{-2} (per operation)

Figure 15.17. *Example of frequency for various types of initiating events*

For the systems that do not function continuously (loading or unloading operations, batch operations, etc.), the occurrence frequency or failure rate must be corrected to account for the actual operating time. This is carried out by multiplying the basic value by the fraction of usage time.

Similarly, the frequencies of initiating events for the human errors are given in relation to the number of operations carried out. We must evaluate the number of operations per period considered, the year in general, to obtain the annual frequency of the error.

EXAMPLE 15.2.– A *batch* production operation takes place 40 times/year and lasts for 2 h. The failure rate of the pump used for the agitation is of $f = 10^{-2}$/year. The frequency to consider in the analysis is $f = 10^{-2} * 40 * 2/8,000$ per year.

15.6.5. *Identification of the safety barriers*

Each protection layer can be classified in relation to its proximity to the source (Figure 15.13), the closest ones having to be the first ones to activate. For a manufacturing process, we will distinguish the following IPL groups:

– procedure design;

– procedure control system;

– alarm and operator's response;

– SISs;

– other barriers for minimizing consequences (containments, safety valves, etc.).

We can add to these the barriers that act at the level of the targets (plant emergency response and community emergency response).

The first phase in the analysis of the IPLs consists of listing the existing barriers and verifying that they are indeed independent. Then, the PFDs for each barrier must be evaluated, while using generic tables (Figure 15.18) [AIC 01], because the LOPA method is a simplified method.

IPL	PFD
Regulation loop	10^{-1}
Safety valve	10^{-2}
Operator's response to an alarm	10^{-1}
Container under pressure	10^{-4} (or better following maintenance)

Figure 15.18. *Several PFD examples*

REMARK 15.2.– The DPPR/SEI2/MM-05-0316 letter from 07/10/05 concerning industrial installations introduces the notion of defense line, equivalent to the notion of IPL of the LOPA method, and classifies them into eight categories, six of which are within the site:

1) design, construction, formation, maintenance, inspection, training;

2) process control system, operator supervision;

3) safety alarms, operateur response;

4) safety programmable logic controller;

5) ultimate safety barriers;

6) emergency plan (internal and external to the site);

7) urbanization control;

8) public information about the emergency plan.

15.6.6. *Calculating the risk level of a scenario*

When the frequency of the occurrence and the severity are determined, the scenario will be assessed at an acceptable or unacceptable level by using the usual approach. Another possibility consists of settling an objective in terms of the probability of the consequences. The table given in Figure 15.19, proposed by Dowell [DOW 98] can be used.

Severity of final consequences	Consequences	Objective for the frequency of the final event	Basis
Minor	Impact limited to the area local to the phenomenon with larger potential consequences if no action is taken	Depends on the cost of protection means in relation to the cost of damages	
Serious	Impact capable of causing serious injury or death on site or outside of the site	10^{-6}	Corporate risk criteria
Major	An impact five times more significant than the 'serious' level	10^{-8}	Two orders of magnitude lower than for the serious level

Figure 15.19. *Objective for the frequency of final events*

Starting from the probability we are concerned with and the actual probability that is assessed for the scenario, it is possible to determine the necessary reduction level and therefore the SIL level that must be added in the barriers. For example, if we must lower the probability of a 10^{-2}, factor, we add an SIL level of 2.

15.6.7. *Example*

Let us once again take the reactor presented in Appendix 7. Let us consider a scenario (Figure 15.20) related to a failure of the cooling system. The planned barriers are as follows:

– IPL1: an automatic system for triggering the sprinkler, controlled by a SIS connected to the sensor TI3307.

– IPL2: a manual system for triggering the sprinkler, with an alarm connected to sensor TI33072 and a valve activated by the operator.

– IPL3: a set of PSV33009 safety valves.

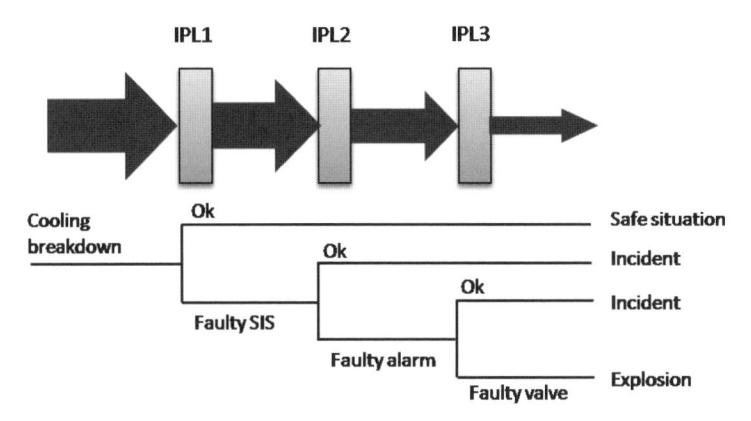

Figure 15.20. *Scenario for the chemical production installation*

Figure 15.21. *LOPA table for the chemical production installation*

The scenario is described in the LOPA table (Figure 15.21). We suppose that its severity level is "serious". The different IPLs are described in the table. We can consider that they are independent. A fault tree allows us to estimate the frequency of the initiating event at 0.1 year^{-1}. The PFD of the IPL2 is 0.1

and of the IPL3 is 10^{-2}. The frequency of the final event is $0.1 \times PFD_1 \times 0.1 \times 10^{-2} = PFD_1 \times 10^{-4}$. The risk has a severity of a high level. By using the table given in Figure 15.19, the target mitigated event likelihood is 10^{-6}. Consequently, the SIS system has an SIL level of 2 so that its PFD_{avg} is 10^{-2}.

15.6.8. Conclusion

The LOPA method described here was developed for the chemical industry. However, its principles are very general and can be used for other applications such as the analysis of emergency management systems. It can also apply to other fields such as medical systems and organizational systems because the barrier analysis principle is very general.

Part 4
Appendices

Appendix 1

Occupational Hazard Checklists

Occupational risks are the risks that affect operators during their professional activity. We can classify them according to the following categories:

– risks related to physicochemical phenomena and machine-related risks;

– risks related to the physical environment and work conditions;

– risks related to the physical activity and the organization of the work.

Tables A1.1–A1.11 list these risks. Each table regards a family of risks and details the representative hazardous situations associated with this risk.

Family	Risks	Dangerous situation
Mechanical Risks	Cutting risks	Use of cutting equipment
		Use of cutting materials
		Use of mechanical cutting machines
		Working in the proximity of cutting elements
	Risk of collapse	Presence of moving elements
		Using a press
		Using crashing equipment (pneumatic drill, sledge hammer, chipping chisel etc.)
	Projection risk	Protection against metallic flash
		Glass protection
		Protection against particles in the eyes
		Protection against mechanical pieces
		Protection against moving mechanical pieces

Table A1.1. *Mechanical risks*

Family	Risks	Dangerous situation
Electrical risks	Risk of electrocution	Intervention on HT live elements (repair)
		Working in proximity to live elements
		Working in proximity to HT lines
		Intervention on the switch box
	Electrification risks	Intervention on the BT live elements (repair)
		Working in proximity to live elements
		Intervention on the switch box
	Risks related to electrostatic charges	Presence of static electricity at the work post

Table A1.2. *Electrical risks*

Family	Risks	Dangerous situation
Chemical risks	Risks related to product stock	Stocking products that have been identified via danger pictogram
		Stocking of domestic products
	Use of chemical products	Use of chemical products
		Working near chemical products
		Decanting chemical prducts
		Explosion with toxic vapors or toxic smoke
		Projection of chemical products
	Anoxia	Sewage network intervention
		Intervention in confined space
		Intervention in the silo
		Nitrogen circuit under pressure

Table A1.3. *Chemical risks*

Family	Risks	Dangerous situation
Risks of falling	Risks of plain falling	Slippery floor
		Cluttered floor(obstacles, electric cables, etc.)
		Distorted floor
		Uneven floor(< 0.25 m)
	Risks of falling from height	Fall > 3m
		Fall between 1 m and 3 m
		Fall < 1 m
		Falling on the stairs
	Risks of falling objects	Use of tools or equipment at great height
		Stocking of material at great height

Table A1.4. *Risks of falling*

Family	Risks	Dangerous situation
Work environment	Risks related to noise	Noise < 80 dBA
		Noise < 87 dBA
		Noise > 87 dBA
		Pulsating noise (klaxon, siren, etc.)
	Risks related to light	Lack of natural light
		Not enough light
		Sunglare
		Poorly orientated light
		Working on a backlit screen
	Risks related to the thermal environment	Significant variation in temperatures (> 10°C)
		Exposure to external heating
		Exposure to high temperatures due to activity (hot)
		Exposure to low temperatures during activity (cold)
		No air conditioning
	Risks related to air quality	Exposure to dust (wood, silica, ferrite or others)
		Exposure to asbestos
		Exposure to bad smells
	Risks related to vibrations	Use of a self-propelling truck
		Use of vibrating machines
		Use of heavy loads
	Risks related to stress and mental load	Production pace
		Service pace
		Delay command
		Information overload
		Relational difficulty
	Risks of verbal agression	Public discontent
		Proximity of the public
		Relational difficulties with the team
	Risks of physical aggression	Public discontent
		Night shifts
		Intervention in an isolated neighborhood

Table A1.5. *Risks related to the environment and working conditions*

Family	Risks	Dangerous situation
Travel related risks	Road risks	Use of service vehicles
		Use of two wheels (bicycle or motorcycle)
		Pedestrian/vehicle collision
	Internal circulation risks	Pedestrian/vehicle collision
		Collision between motorized vehicles

Table A1.6. *Road-related risks*

324   Risk Analysis

Family	Risks	Dangerous situation
Risk related to manual handling and work postures	Risk related to manual handling	Carrying loads> 50 Kg
		Carrying loads > 25 Kg
		Carrying loads < 25 Kg
		Carrying overwhelming loads
		Carrying loads for a longer period of time
		Difficulty in gripping loads
		Traveling > 10 m with loads
	Risk related to gestures and work postures	Prolonged upright, static position
		Prolonged sitting position
		Prolonged kneeling position
		Working with arms postioned above the chest
		Work requiring twists, flexing or twisting of the torso
		Poor ergonomics at the workplace (e.g. screen too high, keyboard too low)
		Floor charging plugs
	Risk related to repetitive gestures	Repetitive strain injury
		Conditioning activities
		Quick sequences of identical gestures

Table A1.7. *Manual handling risks*

Family	Risks	Dangerous situation
Risks related to mechanical handling	Risks of the loads falling	Holding the loads with the aid of slings
		Holding the loads by existing lines
		Holding loads that are heavier than the maximum capacity of the maintenance equipment
		Moving under the suspended loads
	Risks of getting crushed by the loads equipment	Holding loads in a restricted space
		Holding loads while maintaining balance
		Working near suspended loads
	Risks of the mechanical handling and lifting	Working near suspended loads handling and lifting equipment
	Risks of shock with the load	Working near suspended loads
		Passage under the suspended loads

Table A1.8. *Mechanical handling risks*

Appendix 1   325

Family	Risks	Dangerous situation
Biological risks	Risks related to exposure to infectious agents	Exposure to lethal viruses (HIV)
		Exposure to viral agents
		Exposure to bacteria
		Contact with fungi
	Risks related to exposure to infectious agents, risks related to exposure to allergen agents (respiratory and dermatological)	Exposure to pollen
		Exposure to urticating products
		Exposure to dust
		Exposure of asthmatic people to air conditioning
		Exposure to cleaning products
		Exposure to specific chemical products
		Exposure to food powders (e.g. flour)
		Exposure to animal hair

Table A1.9. *Biological risks*

Family	Risks	Dangerous situation
Risks of explosion/fire/heat	Fire risks of electrical origin	Electrical fault of the installation
		Non-conformity of the electrical installation
		Fault of a turned-on electrical device
		Overcharge of electrical plugs
	Fire risks of chemical origin	Stocking of inflammable objects
		Use of inflammable products
		Mixture of incompatible products
	Fire risks via hot spots	Welding
		Projection of incandescent elements
		Overheating of a piece of equipment
		Waste-disposal fire
	Burning risks	Working near hot spots
		Contact with a hot surface
		Use of hot tools
		Projection of incandescent elements
	Explosion risks	Stocking of explosive materials
		Mixture of chemical products
	Risks related to devices and equipment under pressure	Use of gas cylinders
		Use of equipment under pressure
		Working near pressure pipe lines
		Use of autoclave units

Table A1.10. *Explosion and fire-related risks*

Family	Risks	Dangerous situation
Various risks	Risks related to animals	Presence of insects (stings)
		Presence of snakes (bites)
		Presence of dogs
	Drowning risks	Working near water storages
		Intervention in water pools/basins
		Intervention near watercourses
		Intervention in sewers
	Burying risks	Intervention in trenches
		Building retaining walls
		Ground collapsing

Table A1.11. *Various risks*

Appendix 2

Causal Tree Analysis

A2.1. Data gathering

The first step in building a causal tree in order to carry out an analysis after the occurrence of an accident is data gathering.

DEFINITION A2.1.– *A fact is defined as an objective, concrete, precise and verifiable piece of information, which can be recorded in an unbiased manner.*

We are mainly interested in the facts that lead to the implementation of the accident. Gathering of facts can be done using either the 5M method or the ITAME method.

	ITAMAMI	5M
Who?	Individual	Craftsmanship
How?	Task	Methods
With what?	Material	Machines, materials
Where?	Medium	Medium

To be as exhaustive as possible, we must examine the set of elements in the work situation and, first and foremost, analyze the unusual facts. From a practical point of view, we must analyze the locations of the accident as early as possible and then explain to other people what we do and why. Then, we must pick out all the facts in random order, without connecting them at this stage. If necessary, we must explain the usual work procedure. It is important to interview all the parties concerned, by asking open and neutral questions, and then verify and regroup that information. Throughout this stage, it is advisable to take notes.

A2.2. Building the tree

This phase can be carried out within the workgroup with the different people affected by the accident. The construction begins with the analysis of the last fact (or of the last facts). Then, for each fact, we must ask the following questions:

– What was needed for that to happen?

– Is it necessary?

– Is it sufficient?

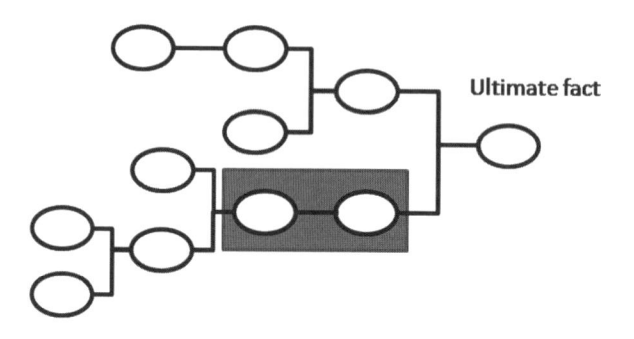

This questioning or examination allows us to identify the causes of a fact that are represented in the graph. Then, we verify whether this cause or this conjunction of causes triggers the fact currently analyzed. Once the tree is built, it serves as the basis for the analysis of possible solutions for preventing this accident in the future.

Appendix 3

A Few Reminders on the Theory of Probability

A3.1. Basic notions

The notion of probability is defined on the basis of a set of outcomes, denoted by Ω, which is the set of all the possible results of a test.

An event A is a subset of Ωs. The probability of an event is denoted by $P(A)$. It is a number between 0 and 1.

DEFINITION A3.1.– *Let $\Psi(\Omega)$ be an algebra of events defined in Ω. A probability in $(\Psi(\Omega), \Omega)$ is, by definition, an application of $\Psi(\Omega)$ to $[0, 1]$:*

$$P : \begin{array}{c} \Psi(\Omega) \to [0, 1] \\ A \to P(A) \end{array}$$

such that:

– $P(\Omega) = 1$

– $P(A \cup B) = P(A) + P(B)$ if $A \cap B = \emptyset$

The probability of an event is denoted by $P(A)$. If $P(A) = 0$, we call the event impossible.

THEOREM A3.1.– (ADDITION RULE OF PROBABILITY).– For any two events, we show that: $P(A \cup B) = P(A) + P(B) - P(A \cap B)$.

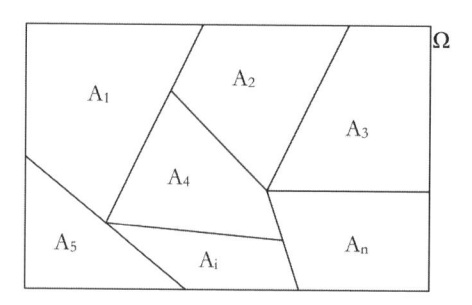

Figure A3.1. *Probability space*

DEFINITION A3.2.– (VENN DIAGRAM).– *A Venn diagram allows us to visually represent the basic properties of the probabilities. In a Venn diagram (Figure A3.2), the sampled space is represented by a rectangle and each event, subset of Ω, is represented by a geometrical form (often a circle) of a surface that is directly proportional to the probability of this event.*

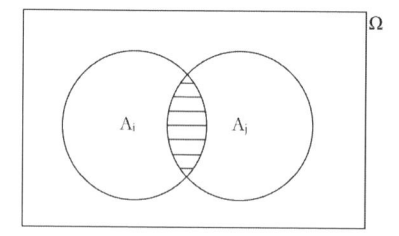

Figure A3.2. *Venn diagram*

DEFINITION A3.3.– (CONDITIONAL PROBABILITY).– *Let us assume that we wish to know the probability of an event A, knowing that an event X has already taken place. We, therefore, wish to know the relative proportion of A relative to X (Figure A3.3). We therefore define the conditional probability of A knowing X thus:*

$$P(A|X) = \frac{P(A \cap X)}{P(X)}$$

We can write:

$$P(A \cap X) = P(A|X)P(X) = P(X|A)P(A)$$

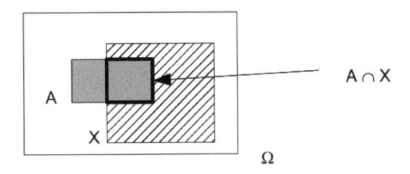

Figure A3.3. *Conditional probability*

If two events are independent, i.e. if the occurrence of one does not give any information on the occurrence of the other, we can write $P(A|B) = P(A)$, which gives the following calculation rule:

$$P(A \cap B) = P(A) \times P(B)$$

When this equality is not verified, the events A and B are said to be independent.

DEFINITION A3.4.– *A random variable is a function defined on the set of possible outcomes, i.e. the set of possible results of a random experience. A real random variable takes its values from the set of real number.*

Examples of random variables:

– The lifetime of a component is a real random variable.

– The variable is defined as being equal to 0 if the system has broken down, or equal to 1 if working properly is a discrete random variable on $\{0,1\}$.

DEFINITION A3.5.– *The distribution function of a real random variable X is defined by the function F of $\mathbb{R} \to [0,1]$ $F(x) = P[X \leq x]$. When F is differentiable, the density of the probability of X is defined by $f(x) = \frac{dF}{dx}(x)$. These functions are such that:*

– $F(x) = \int_{-\infty}^{x} f(u)du$;

– $\int_{-\infty}^{+\infty} f(u)du = 1$.

A3.2. Useful distribution

A3.2.1. *Exponential law*

This distribution allows us to represent the failure rate of electronic systems. It is very often used if no more detailed information is available. It is a law which is defined by the probability density function $f(x) = \lambda e^{-\lambda t}$ if $x \geq 0$ and the distribution function $F(x) = 1 - e^{-\lambda t}$ if $x \geq 0$.

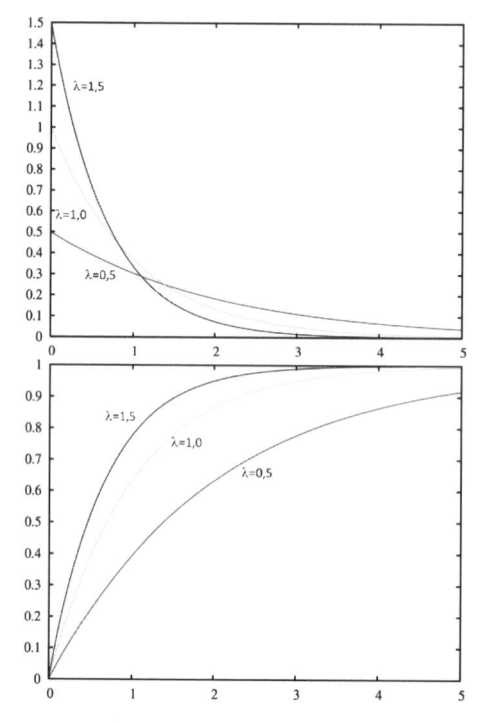

Figure A3.4. *Exponential law*

A3.2.2. *Normal distribution*

The normal distribution is defined by the following probability density function and distribution function:

- $f(x) = \frac{1}{\sigma\sqrt{2\pi}} \exp\left(-\frac{(x-\mu)^2}{2\sigma^2}\right)$;
- $F(x) = \frac{1}{2}\left(1 + \text{erf}\left(\frac{x-\mu}{\sigma\sqrt{2}}\right)\right)$.

A3.2.3. *The Weibull distribution*

The Weibull distribution allows us to represent the fault probabilities of mechanical systems. It is defined by the following probability density function and distribution function:

- $f(x) = \frac{k}{\lambda} \left(\frac{x-\theta}{\lambda} \right)^{k-1} e^{-\left(\frac{x-\theta}{\lambda} \right)^k}$;
- $F(x) = 1 - e^{-\left(\frac{x-\theta}{\lambda} \right)^k}$.

In the above equations, $k > 0$ is the shape parameter, λ is the scale parameter and θ is the location parameter of the distribution.

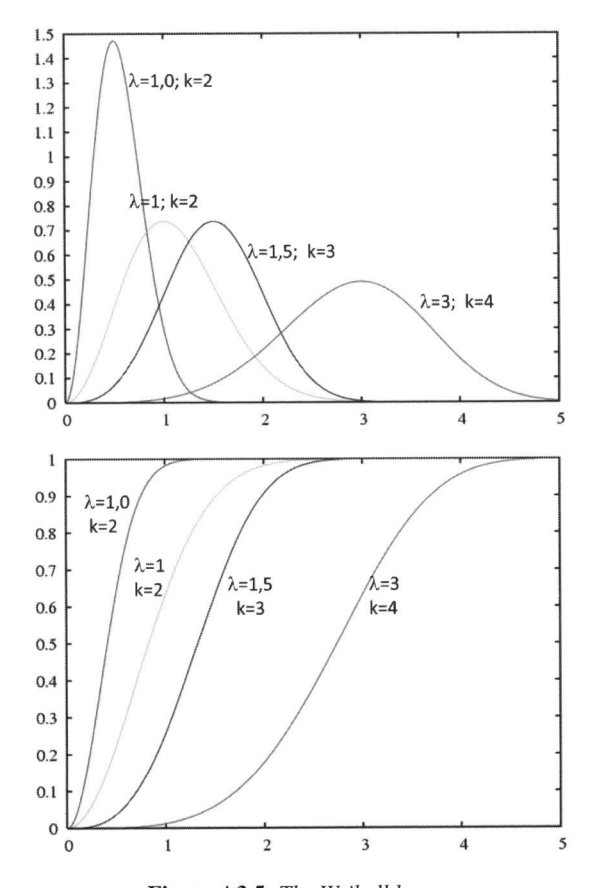

Figure A3.5. *The Weibull law*

334   Risk Analysis

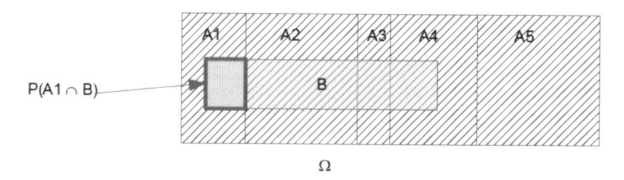

Figure A3.6. *The Bayes theorem*

A3.3. The Bayes theorem

The Bayes theorem allows us to update the probability of an event, considering the observations. Let us consider an Ω partition, denoted $A_1, ... A_n$, and B an event of Ω. We have the following relation:

$$P(A_i|B) = \frac{P(B|A_i).P(A_i)}{\sum_{j=1}^{n} P(B|A_j).P(A_j)}$$

Appendix 4

Useful Notions in Reliability Theory

A4.1. General notions

A4.1.1. *Definition*

DEFINITION A4.1.– *(standards X60-500): reliability is the ability of an entity to perform a certain required function, in some given conditions, during a given period of time. It is measured by the probability that the entity performs the required function during the time interval* $[0, t]$, *assuming that it functions normally at time* $t = 0$:

$$R(t) = P\{E \text{ good functioning at } [0, t]\} = P[T > t]$$

where T *is the time when the entity fails.*

DEFINITION A4.2.–*The mean time between failures (MTBF) is defined as the mathematical average of the time between the failures of a repairable system. We also define the following terms (Figure A4.1):*

- *the MTTF:* mean time to failure;
- *the MDT:* mean down time;
- *the MTTR:* mean time to repair;
- *the MUT:* mean up time.

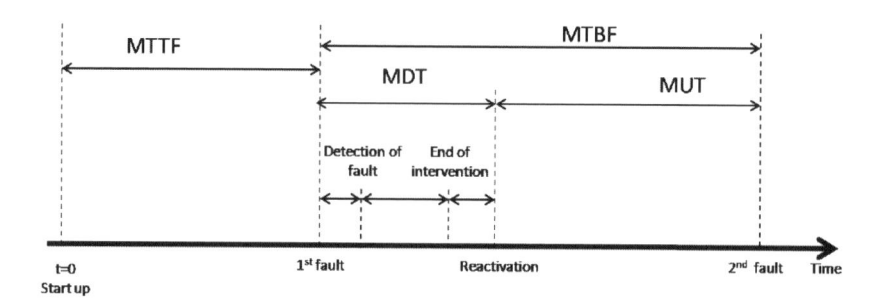

Figure A4.1. *Durations used in reliability*

DEFINITION A4.3.– *The failure rate $\Lambda(t)$ is defined as the probability that the entity fails between t and $t + dt$, knowing that it was functioning at instant t.*

The failure rate and the reliability function (the probability that the component functions up to instant t) are connected by the following relations:

$$R(t) = e^{-\int_0^t \Lambda(u)du} \text{ and } \Lambda(t) = -\frac{1}{R(t)}\frac{dR(t)}{dt}$$

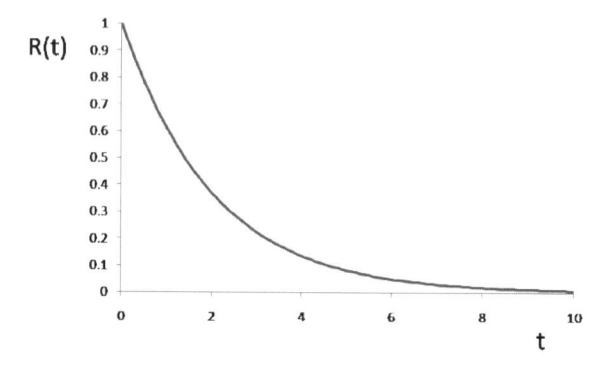

Figure A4.2. *Example of reliability function*

A4.1.2. Bathtub curve

The plot of the failure rate as a function of time usually looks like a bath curve (Figure A4.3). We can distinguish between three periods:

– The youth of the product: the failures are due to manufacturing failures or other rapidly evolving phenomena. The failure rate decreases with lifetime. This period has a variable duration depending on the product. It ranges from several hours to several hundreds of hours.

– The second presents a fairly constant failure rate: it corresponds to the appearance of failures due to very different causes. This period corresponds to the useful lifetime. Its duration ranges from several thousands of hours (for the mechanical pieces) to several hundreds of thousands of hours (for the electronic components).

– The last one is characterized by an increasing failure rate. It corresponds to the appearance of failures due to wear or strain (end of lifetime).

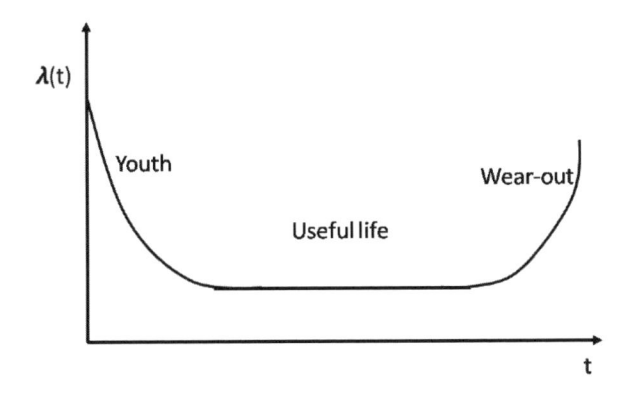

Figure A4.3. *Variation of the failure rate depending on lifetime*

This curve allows us to justify the choice of a constant failure rate, during the useful lifetime of the product.

A4.2. Assessment of the failure rate of a component

The failure rate of the existing components can be obtained by using data based on the observation of a set of these components. We define the total duration of the functioning of the set of these components by T_c, which is equal to the sum of the durations of good functioning of all the entities, and the number of faults observed by *Ndf*. We can estimate the failure rate, which is supposed to be constant, using the following formula:

$$\hat{\lambda} = \frac{Ndf}{T_c}$$

Similarly, the reliability law can be interpreted as the ratio of the components and their good functioning:

$$\hat{R}(t) = \frac{N_{ok}(t)}{N_0}$$

Figure A4.4 illustrates these measurements for a sample of lamps.

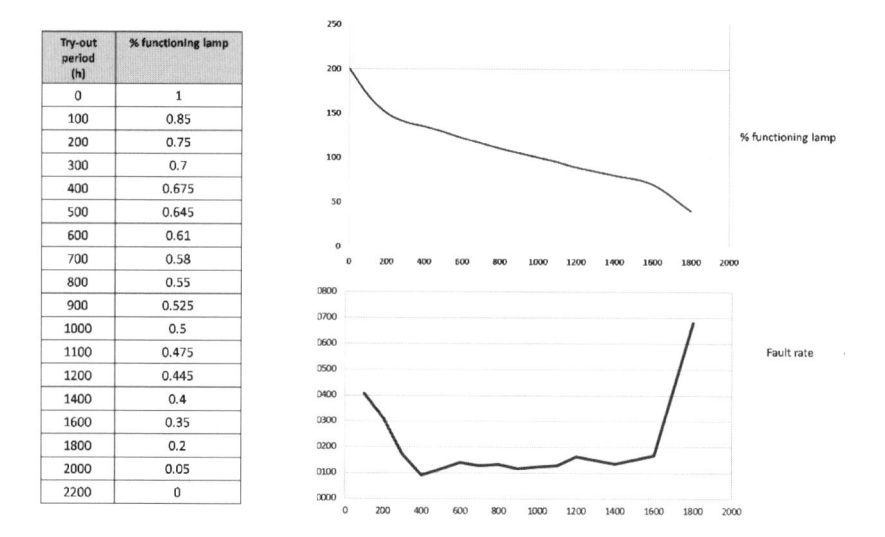

Try-out period (h)	% functioning lamp
0	1
100	0.85
200	0.75
300	0.7
400	0.675
500	0.645
600	0.61
700	0.58
800	0.55
900	0.525
1000	0.5
1100	0.475
1200	0.445
1400	0.4
1600	0.35
1800	0.2
2000	0.05
2200	0

Figure A4.4. *Example of an experimental determination of the failure rate*

A4.3. Assessment of the failure probability of a device

For electronic components, we most often use a constant failure rate distribution. The failure probability is equal to $1 - R(t)$ if:

$$P(t) = 1 - e^{-\lambda t} \quad \text{and} \quad \text{MTBF} = \frac{1}{\lambda}$$

For mechanical systems, we use the Weibull distribution:

$$P(t) = 1 - R(t) = 1 - e^{-(\frac{t-\theta}{\lambda})^k} \quad \text{and} \quad \text{MTBF} = \int_0^\infty R(t)$$

A4.4. Reliability of a set of components and reliability blocks

A4.4.1. *Series components*

The reliability diagram presented in Figure A4.5 expresses the fact that the system is faulty if only one of them is faulty. The probability for normal functioning of the system can be written as follows:

$$P_{ok} = P(E_1 \text{ and } E_2 \text{ ..and } E_n) = P(E_1).\ldots.P(E_n)$$

if we assume the components are independent. We therefore have:

$$R(t) = R_1(t) \ldots R_n(t)$$

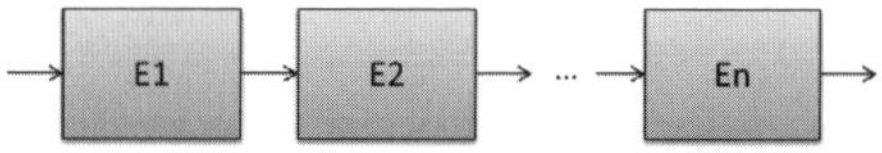

Figure A4.5. *Reliability block diagram for the series systems*

By using the expression $R(t) = e^{-\int_0^t \Lambda(u)du}$, we obtain:

$$R(t) = e^{-\int_0^t \{\sum_{i=1}^n \Lambda_i(u)\}du}$$

The failure rate is the sum of the failure rates: $\Lambda(t) = \sum_{i=1}^n \Lambda_i(u)$.

In the case of a distribution with a constant rate, we obtain a very useful relation:

$$\Lambda = \frac{1}{MTTF} = \sum_{i=1}^n \frac{1}{MTTF_i}$$

A4.4.2. *Parallel components*

The reliability diagram presented in Figure A4.6 expresses the fact that the system is faulty if all of them are faulty. The probability for normal functioning of the system can be written as follows:

$$P_{ok} = P(E_1 \text{ or } E_2 \text{ ..or } E_n)$$

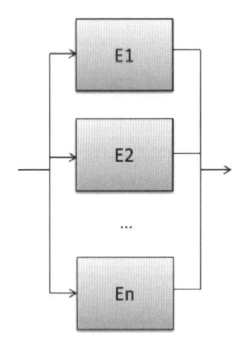

Figure A4.6. *Reliability block diagram for parallel systems*

If we assume the components are independent, the fault probability is therefore:

$$P_{\overline{ok}} = P(E_{1def} \text{ and } E_{2def} \text{ ..and } E_{ndef}) = P_{def}(E_1) \ldots P_{def}(E_n)$$

We, therefore, obtain $1 - R(t) = (1 - R_1(t)) \ldots (1 - R_n(t))$. Using the expression $R(t) = e^{-\int_0^t \Lambda(u)du}$, we obtain:

$$R(t) = 1 - \prod_{i=1}^{n}\left(1 - e^{-\int_0^t \Lambda_i(u)du}\right)$$

For example, in the case of two components with a constant failure rate:

$$R(t) = e^{-\lambda_1 t} + e^{-\lambda_2 t} + e^{-(\lambda_1+\lambda_2)t}$$

More details can be found in [SMI 00].

Appendix 5

Data Sources for Reliability

A5.1. Organizations

– INERIS, the National Institute of Industrial Environment and Risks (Institut National de l'Environnement Industriel et des Risques, www.ineris.fr), is an organization whose mission is to assess and prevent accident-related risks or chronic risks for people and the environment, risks related to industrial installations, chemical substances and underground exploitations. It publishes several reference guides.

– INRS, the National Institute of Research in Safety (Institut National de Recherche en Sécurité, www.inrs.fr) is a leading organization for the prevention of professional risks (work accidents, occupational diseases, etc.). It also publishes several reference guides.

– IMDR, the Institute for Risk Control (Institut pour la Maîtrise des risques, www.imdr.fr), is a leading association, whose aim is to help businesses and public organizations undergo preventive processes in order to correctly identify, assess, quantify, put into hierarchical order, control and manage undesirable events. It manages several work groups in the field of risk analysis.

– GESIP, Research Group in the field of Oil and Chemical Industries (Groupe d'étude de Sécurité des Industries Pétrolières et Chimiques, www.gesip.com), is a group of manufacturers whose mission is to promote experience feedback, train personnel, bring responses to technical evolutions and promote security.

– ICSI, Institute for Culture of Industrial Safety (Institut pour une Culture de Sécurité Industrielle, www.icsi-eu.org), is a meeting point for discussions. Its aim is to develop the culture of safety. It hosts discussion groups and publishes guides.

– UBA, Umwelt Bundesamt (www.umweltbundesamt.de/index-e.htm), is the German Federal Environment Agency whose mission is to bring scientific expertise for implementing laws, gather and disseminate data on the environment, and inform the public of what is at stake from an environmental viewpoint.

– HSE, Health and Safety Executive, (www.hse.gov.uk), is a UK-based organization, which helps the Health and Safety Commission (HSC) to guarantee that risks to the health and safety of people in the workplace are appropriately kept under control in Great Britain. It has particularly defined the As Low As Reasonably Practicable (ALARP) principle.

– TNO, Nederlandse Organisatie voor Toegepast Natuurwetenschappelijk Onderzoek (Dutch Organisation for Applied Scientific Research), www.tno.nl/index.cfm?Taal=2, is the known author of *colored books* (*yellow, green, purple, red*).

– DNV, Det Norske Veritas, Norway (www.dnv.com), is an independent foundation, initially created for the inspection of ships, whose objective is to preserve life, property and the environment and whose core business is to identify, assess and advise in the field of risk management.

– JRC, Joint Research Center (http://ec.europa.eu/dgs/jrc/), is a research support organization attached to the European Commission. This organization provides scientific opinion and technical knowledge. It is divided into several institutes, more particularly:

- Institute for the Protection and Security of the Citizen (IPSC), which hosts the Major Accident Hazards Bureau (MAHB);

- Institute for Health and Consumer Protection (IHCP);

- Institute for Prospective Technological Studies (IPTS).

– ESReDA, European Safety, Reliability and Data Association (www.esreda.org), is a European association which provides a forum for information exchange, data and current research exchange on safety and reliability.

– EPA, Environmental Protection Agency, (www.epa.gov), is an American organization whose mission is to protect the health of people and the environment in the United States, by being of assistance to public powers.

– CCPS, Center for Chemical Process Safety, created by the American Institute of Chemical Engineers (www.aiche.org/ccps), is an American organization that is concerned with the safety of procedures in the chemical, pharmaceutical and oil industries. The CCPS gathers together the manufacturers, governmental organizations, consultants, universities and insurers, and publishes several works and recommendations.

– FEMA, Federal Emergency Management Agency, (www.fema.gov), is an American organization whose mission is to assist citizens and emergency managers in the implementation of a plan of action in emergency situations.

– OSHA, Occupational Safety and Health Administration (www.osha.gov), is an agency of the US Department of Labor, which regulates health and safety in the workplace.

– HSS, the Office of Health, Safety and Security (www.hss.doe.gov), is an agency of the US Department of Energy, which groups together the aspects related to health and safety.

– PHSMA, Pipeline and Hazardous Materials Safety Administration (http://phmsa.dot.gov), is an agency of the US Department of Transportation, which groups together the aspects related to safety.

– EGIG, European Gas Pipeline Incident Data Group (EGIG) (www.egig.eu), is a European organization, which collects data on accidental leakages in pipelines in Europe.

A5.2. Databases

A5.2.1. *Accidents*

– ARIA, *Analysis, Research and Information on Accidents*, exploited by the Office of Risk Analysis and Industrial Pollution (ORAIP) (www.aria.developpement-durable.gouv.fr/index.html).

– MARS, *Major Accident Reporting System*, maintained by the MAHB and is related to the European directive Seveso II on major accidents involving highly hazardous substances (emars.jrc.ec.europa.eu).

– ZEMA, maintained by the German Environment Agency (www.infosis.uba.de).

– FACTS, *Failure and Accident Technical Information System*, run by Unified Industrial & Harbour Fire Department of Rotterdam Rozenburg (www.factsonline.nl).

– HCR, *Offshore Hydrocarbon Release*, managed by the HSE (www.hse.gov.uk/offshore/hydrocarbon.htm).

– PSID, *Process Safety Incident Database*, managed by the CCPS (http://www.aiche.org/ccps/resources/psid).

– ECCAIRS, *European Co-ordination Centre for Aviation Incident Reporting Systems Aviation Accident Database & Synopses*, managed by the National Transportation Safety Board, United States (www.ntsb.gov/default.htm).

– RMP*Info, managed by the EPA (www.epa.gov/oem/content/rmp/).

– JSP, gathers data on accidents or incidents in Japan, (http://shippai.jst.go.jp/en).

– EGIG *Database, European Gas Pipeline Incident Database*, managed by the EGIG, (www.egig.eu).

A5.2.2. Reliability data

– Frequency of the accident initiating events, the notebooks of industrial safety, 09/2009, ICSI (www.icsi-eu.org/).

– *Failure Rate and Event Data for use within Risk Assessments*, HSE, *Health and Safety, Executive*, 06/2012 (www.hse.gov.uk).

– US MIL HANDBOOK 217, *Reliability Prediction of Electronic Equipment*, US Department of Defense.

– *Handbook of Reliability Data for Electronic Components* RDF2000, CNET, France Telecom, National Telecommunications Research Center.

– HRD: *British Handbook of Reliability Data for Components Used in Telecommunications Systems* (HRD5, last issue).

– *Guidelines for Process Equipment Reliability Data with Data Tables*, CCPS, Center for Chemical Process Safety of the American Institute of Chemical Engineers, ISBN 0-8169-042207.

– *Guidelines for Quantitative Risk Assessment, Purple Book*, CPR18E, Sdu, 1999.

– OREDA , *Offshore Reliability Data* (2009), initially managed by the SINTEF, Norway (www.oreda.com).

– PERD, *Process Equipment Reliability Database*, managed by the CCPS (www.aiche.org/process-equipment-reliability-database-perd).

– PDS, methodological guide and data gathering for the reliability analysis of the SIS, managed by the SINTEF (www.sintef.no/pds).

– EIREDA, *European Industry Reliability Data Bank*, European reliability database for industrial materials (1998), managed by the ESReDA.

– IEEE *Guide to the Collection and Presentation of Electrical, Electronic, Sensing Component, and Mechanical Equipment Reliability Data for Nuclear-Power Generation Stations*, IEEE Standard 500-1984.

– *Nonelectronic Parts Reliability Data* 1991, carried out by William Denson, Greg Chandler, William Crowell and Rick Wanner, Reliability AnalysisCenter, (www.dtic.mil/dtic).

– *Reactor Safety Study – An Assessment of Accident Risks in U.S. Commercial Nuclear Power Plants,* NASH-1400 (NUREG-75/014), *Appendix III –Failure Data, October* 1975.

– *An Introduction to Machinery Reliability Assessment*; Tables 4.2, 4.3 and 4.5; Figure 4.8; 1990; Heinz P. Bloch and Fred K. Geitner; Van NostrandReinhold; New York, NY.

– *Handbook on Human Reliability Analysis with Emphasis on NuclearPower Plant Applications*; NUREG/CR-1278; October 1980; A.D. Swain and H.E. Guttmann; U. S. Nuclear Regulatory Commission.

Appendix 6

A Few Approaches for System Modelling

A6.1. The Lemoigne system modeling

Systemic differs from other modeling methods because it develops a holistic approach[1]. Complex systems are no longer considered as the sum of their parts, as per the analytical method: they form an inseparable whole, and allow for the emergence of certain new properties [SIM 96]. A general model of the systems was proposed by Lemoigne [ERI 97] (Figure A6.1). Even though it is difficult to use *per se*, it has inspired a significant number of modeling methods for risk analysis, i.e. the so-called systemic approaches, which analyze the systems globally, as they behave in the normal (or abnormal) interactions they might have with the exterior. Moreover, we note that the emergency property implies that the analytical study of the elements of a system does not allow for a good understanding of global dysfunctions and we, therefore, need a systemic overview if we wish to conduct a good analysis.

A6.2. Representation methods of the standard procedure ISO 9001

The notion of process [BRA 03] is used in the field of quality control and is defined by the standard ISO 9001:2000. A process is defined as a system of

---

[1] Theory or viewpoint consisting of considering phenomena in their totality.

activities that uses resources (personnel, equipment, materials, information) in order to transform input elements into output elements. In order to describe a process (Figure A6.2), we must describe:

– the activities and the succession of activities involved, which can be described as a task diagram;

– the input elements;

– the output elements;

– the parties and resources;

– different parameters such as delays, the interfaces between activities, the control and measurement elements.

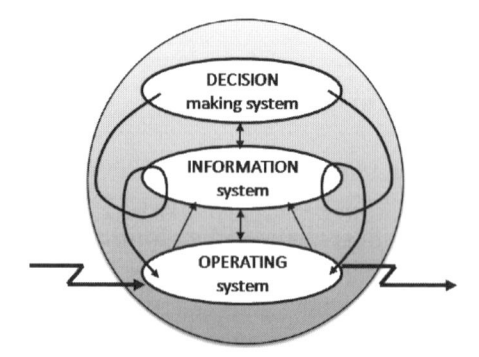

Figure A6.1. *Classical operation–information–decision model*

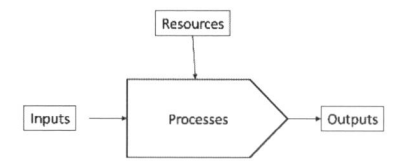

Figure A6.2. *Process model*

The process-oriented approach can be applied hierarchically (Figure A6.3). The use of a process-oriented approach can lead to elaborating a process map of the processes that allow us to represent the organization via connections between the different processes.

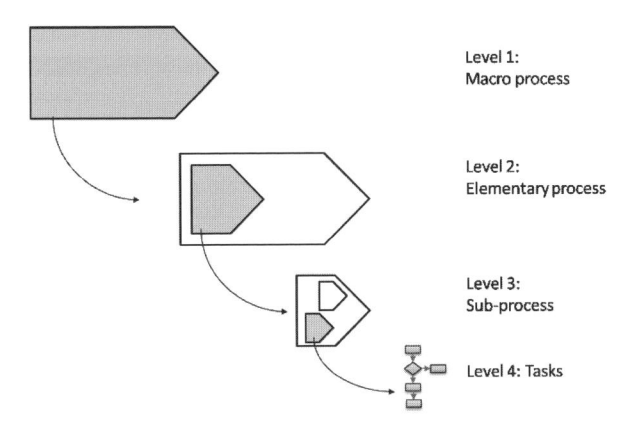

Figure A6.3. *Hierarchical model of a process*

A6.3. Functional analysis methods IDEF0, SADT and FAST

The IDEF0 method is from the IDEF family (*integration definition language*) which was developed in the 1970s as an initiative of the US Army. The IDEF0 method was formalized by D.T. Ross [ROS 77], who has created a society, Softech, to market this method under the name SADT.

The IDEF0 model is a tree of diagrams (Figure A6.4), which includes functions, text and a glossary (Figure A6.5). The basic model is the model representing a function. It is represented by a box with arrows describing the inputs, the outputs, the mechanisms and the controls:

– An input is an element transformed or consumed by the function so as to produce the outputs. This is not an event that activates the function.

– An output is an element produced by the function.

– A control is a piece of information that is used by the function in order to determine its behavior.

– A mechanism or medium is an element needed for the execution of the function (hardware, software, actors, etc.).

A function can have any number of inputs or outputs. They are represented by labels on the arrows (and not by a box).

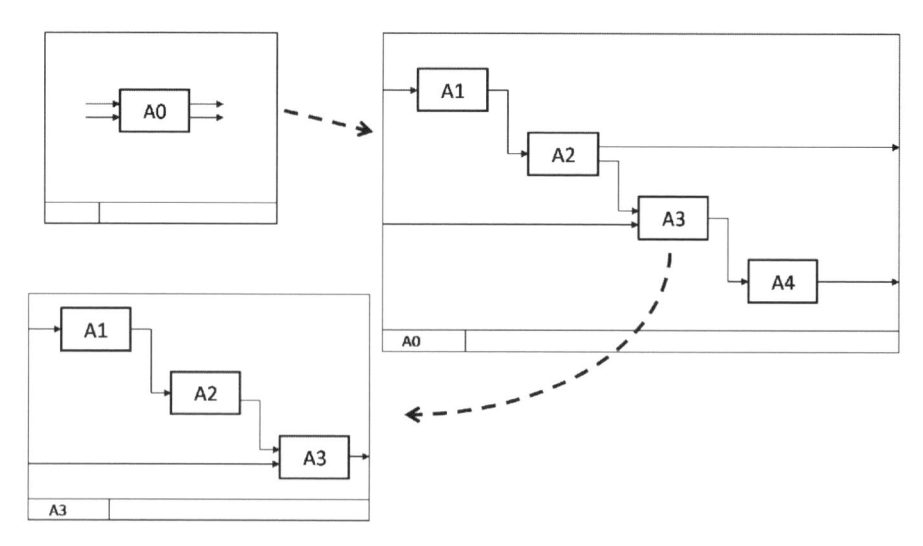

Figure A6.4. *Hierarchical IDEF0 diagrams*

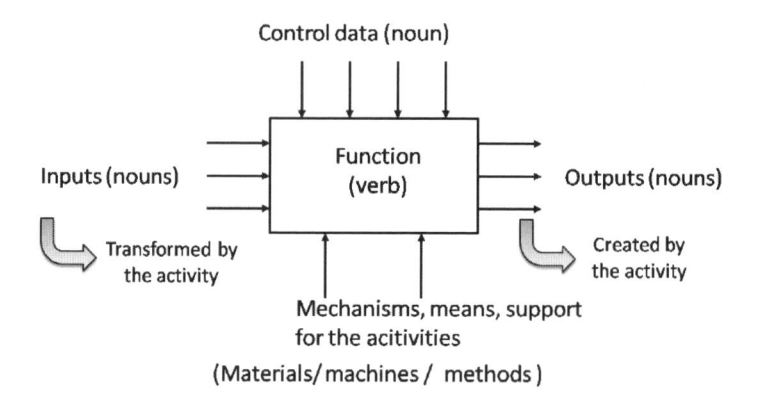

Figure A6.5. *IDEF0 functioning model*

REMARK A6.1.– A functional analysis system technique (FAST) diagram presents a functional decomposition that is different from the SADT method. The FAST diagram is built from left to right, using a logic that goes from the "why?" to the "how?". The NF EN 12973 standard (value-based management) describes the FAST diagram as one of the most common methods of functional analysis. It is, however, used very little for risk analysis. The method is based on an interrogative technique:

– Why?: why must a function be ensured? We answer this question by reading the diagram from right to left, and we move on to higher level functions;

– How?: how must this function be ensured? We then decompose the function, and we can read the answer to the question by going through the diagram from left to right;

– When?: when must this function be ensured?

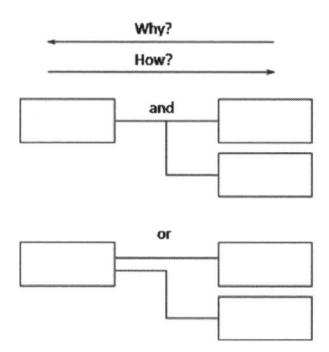

Figure A6.6. *FAST diagram*

A6.4. UML and SYSML methods

The unified modeling language (UML) method was first developed in the IT sector. It proposes 13 diagrams allowing us to describe the structural elements and the behavior of a computer system, while considering the interactions with the user. The UML 2.2 language defines 14 types of diagrams, categorized as follows:

– structural diagrams that describe the structure of the system, such as class diagram, which describes the attributes of a class representing a type of object;

– behavioral diagrams that describe the behavior of the system, such as activity diagram, which describes how the system must evolve.

The UML approach was adapted to the modeling of physical systems; the name of this method is SYSML [WEI 08]. It takes a subset of ten UML diagrams, some of them in an altered version, and adds two new other diagrams (Figure A6.7). These modifications facilitate the representation of physical systems by using notions such as *blocks, item flows, value properties, allocations, requirements, parametrics and continuous flows.*

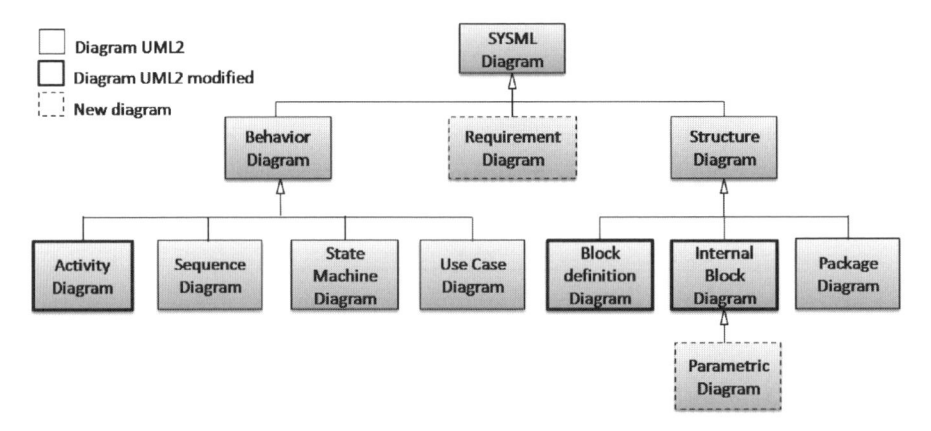

Figure A6.7. *SYSML diagrams*

This approach is suitable for systems specification or simulation modeling, but is relatively cumbersome for risk analysis as a large number of its elements are unnecessary.

A6.5. FIS structural-functioning method

The FIS model is composed of three views [FLA 08, KAR 10]:

– the structural model (SysFis);

– the dysfunctional model (DysFis);

– the evolution model (SimFis).

The basic element of the representation is the process or system, made up of functions, physical resources and input/output gates. A process or system is seen as an organized set of activities that uses resources (staff, equipment, hardware and machines, raw material and information) for transforming the inputs into outputs. Each process is described internally by the following:

– The resources that are made up of hardware elements (machines and material), informational, human and organizational, allow us to carry on the activities of the process. The methods (procedures, operational modes, programs, etc.) are part of the resources. The resources can be characterized by a set of variables.

– The functions or activities of the system that are defined as being the role of a set of resources expressed in terms of purpose. It is a type of dematerialization of a set of entities that indicate what it can do. It represents a specification of the behavior of the system. A function can be active or non-active at any given instant. Its behavior is characterized by a set of variables and, if needed, by a behavioral model.

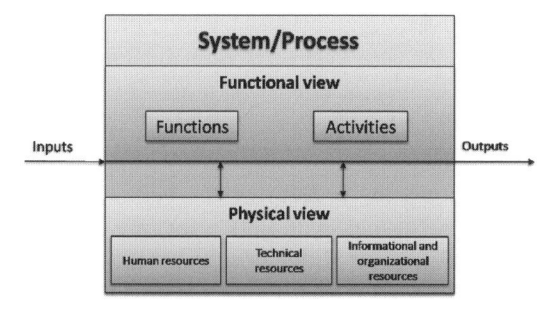

Figure A6.8. *SysFis model*

A function can be connected to resources in several ways:

– It can consume the resources, use their input state or use them to implement itself: the resources are inputs in the function.

– It can generate or act upon them: the resources are then "outputs" for the function.

The dysfunctional model is represented by an event graph. The model classifies the events into the following types:

– events that represent the undesirable hazardous events;

– events that represent the fault modes;

– events that represent the variable deviations;

– events that represent degradations;

– intermediary events.

These events are attached to the elements of the model (systems, functions and resources) and they allow us to structure the analysis (Figure A6.9). Each event has one or several barriers attached to it.

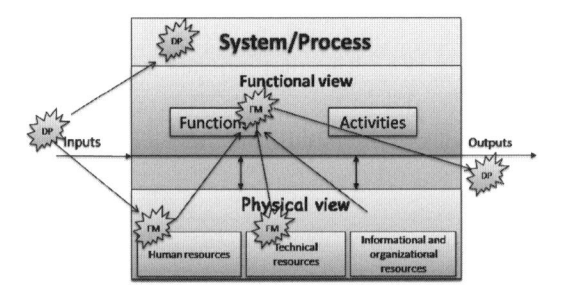

Figure A6.9. *Dysfis model*

The SimFis view allows us to describe the behavior of the functions via a dynamic model with a limited state and a set of relations describing how the input resources are modified by the function, or transformed into output resources; this part of the model allows us to simulate the system in degraded mode.

Appendix 7

Case Study: Chemical Process

A7.1. Basic installation

The system that we are concerned with here is shown in Figure A7.1. The R33030 reactor is used in *batch* mode (reactor closed) to run a chemical reaction in order to produce a product C from two reactives A and B. The reactor is loaded with a volume equivalent to the two products A and B, which triggers an exothermal reaction. The temperature rises up to 120°C, a value where it is regulated with industrial water. Over 100°C, product C exists in the headspace of the reactor as toxic gas. Product B also emanates toxic vapors.

At the end of the reaction, the mixture is completely transformed if A and B are in stoichiometric proportions. It is transferred toward the rest of the chemical process by opening the valve XV33021.

The exchanger E33040 receives cold industrial water (between 3°C and 15°C) as an input which is used to cool down the content of reactor R33030 using a double jacket. The temperature of the exchanger is measured by the sensor TI33061. The pressure in the reactive medium is less than six bars when the temperature is controlled at 120°C. The PSV33009 valve limits the pressure to 10 bars. In case of overpressure, the exhausted gases are cleaned by a scrubber.

The pumps and the motor of the agitator are fed by a line of 380 V. They are controlled via a relay (Figure A7.2). The regulators, sensors and actuators are powered with 48 V. The information is received in the control room (where the programmable logic controller (PLC) and regulators are found) by a fieldbus, that is, a computer bus that chains together the devices one by one, in a series. The regulation of the flow of cold water is carried out by a regulator moved into the equipment room. The same basic process control system (BPCS) manages the loading: for a given quantity of the product C that must be manufactured, it fills up products A and B.

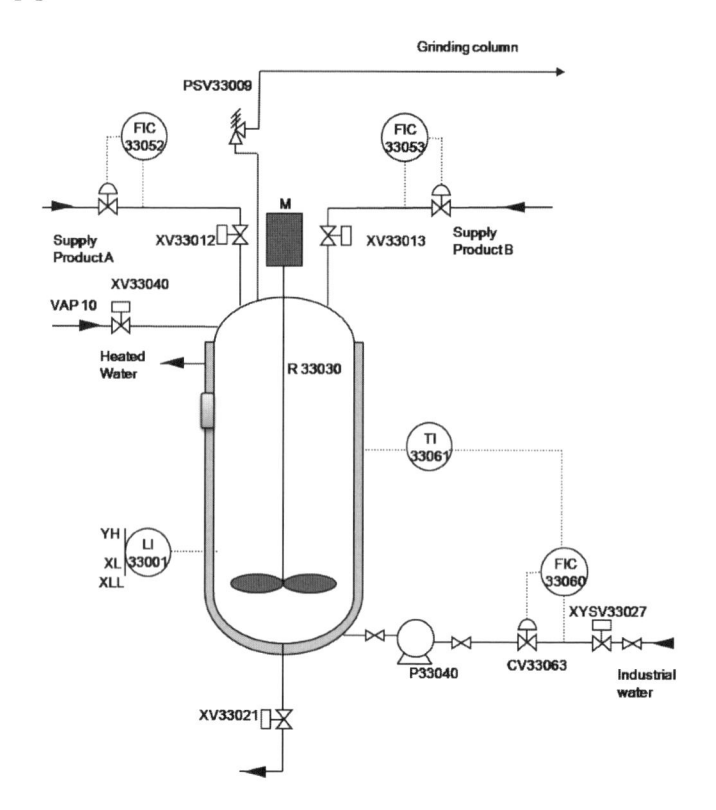

Figure A7.1. *Chemical process*

The reactor is operated in two 8-hr shifts and carries out two production operations per day. The operating mode is as follows:

– checking that the reactor is empty; emptying it if needed;

– vapor cleaning (via line VAP10);

– launching water circulation, temperature regulation and agitation;

– loading:

- the operator enters the quantity of the product that must be produced;

- the BPCS opens the valve for product A to load product A;

- the BPCS opens the valve for product B to load product B;

– waiting while supervising the reaction;

– emptying the reactor.

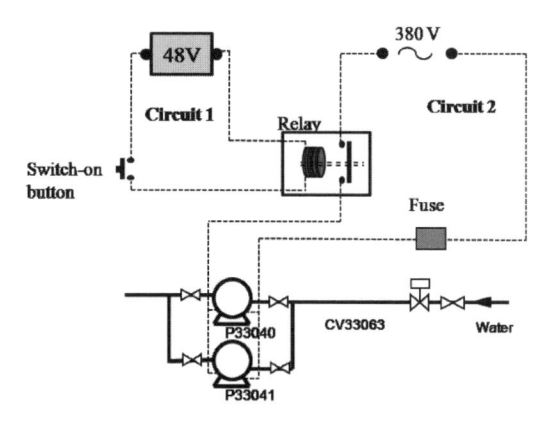

Figure A7.2. *Pump control system*

The reactor is housed in a manufacturing site located south of a small town.

The main risks associated with this installation are those related to:

– a toxic leakage, through a joint, or a leaky valve of products B or C, due to corrosion;

– an explosion in the case of a rise in pressure and the blocking of the valve, if a thermal runaway occurs. This explosion leads to a release of toxic products;

– an electric shock, on the parts supplied with 380 V, essentially during the maintenance phases.

Overpressure is due to a thermal runaway of the reaction, caused by a rise in temperature. This can be due to a breakdown in temperature regulation, or insufficient water, or when the agitator breaks down, or if an overload takes place.

The rise in pressure is limited by the mechanical safety valve PSV33009. If this gets blocked, it would entail the explosion of the reactor and a release of toxic products.

The design of the cooling system was carried out for climatic conditions established on the basis of the history of the area where the site is situated. However, in the case of a heat wave, when air temperature is very high, water temperature rises and the flow is very low, the cooling system could prove to be insufficient. These conditions happen all at once less than once every few centuries however, in light of the experience we have accumulated so far.

Another source of problems for the cooling system comes from excessive loading of the reactor. An excessive loading due to human error (instruction error) can result in the inability of the cooling system to maintain the temperature at an acceptable level. The eventuality of such an error is assessed as being less than once every thousand loadings.

A7.2. Improvement project

An improved version of the process is presented in Figure A7.3. The modifications proposed are the following:

– the cooling circuit pump is doubled, a similar P33041 pump is added;

– the valve is doubled;

– a sprinkling system is added, associated with an independent SIS.

In case of an overheating of the installation, a sprinkling system is activated by a safety PLC. The temperature sensor, the chain of information processing and the opening valve of industrial cold water are independent of those of the basic process control system. The water comes from a tank situated on the riverside and from another tank of 5 m^3 located high enough to allow the natural flow of water. This quantity is enough to stop the reaction. Moreover, the pump for the cooling system was doubled and so was the valve.

The sprinkling can be launched directly by the operator if needed. The detection can be done with the driving information or on an alarm when the temperature gets too high, activated from a third temperature sensor, TI 33072.

With means of prevention

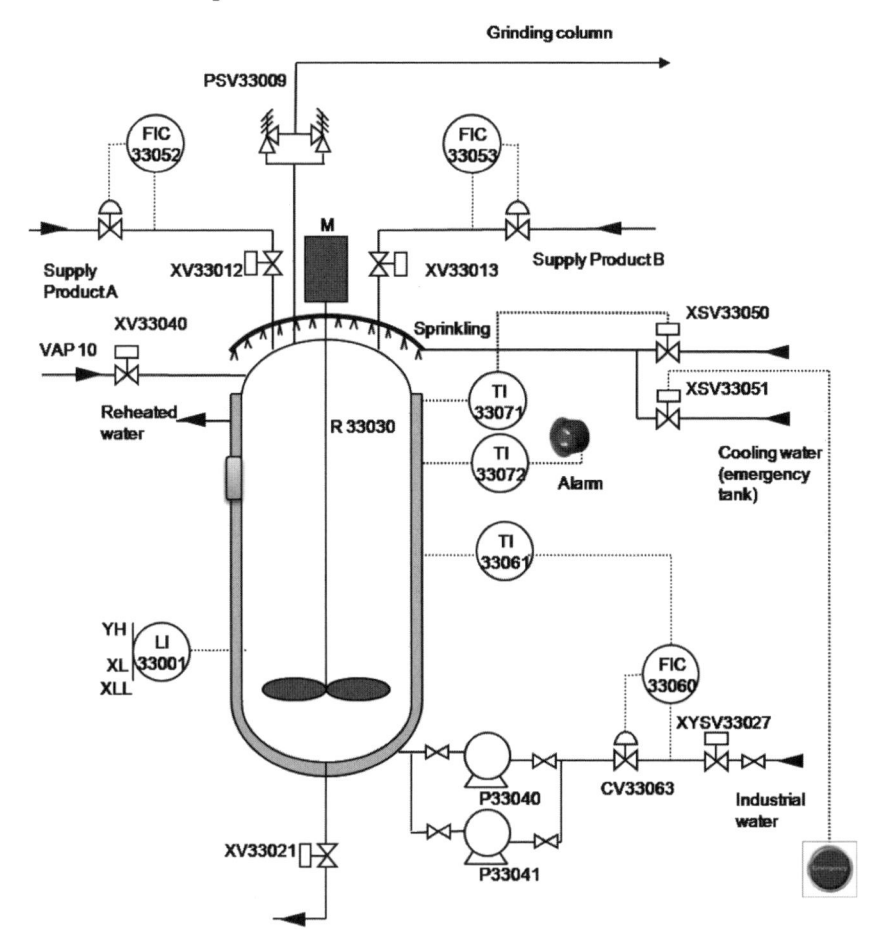

Figure A7.3. *Chemical process with safety barriers*

In case of alarm, the procedure applied by the operator is the following:

– verifying that the sprinkler is functioning (a visual check from the control room is sufficient);

– manually activating the sprinkler pump XSV33051.

Reliability data:

– fault in the supply of cooling water: $< 10^{-2}$/year;

– joint fault: 10^{-7}/year;

– piping failure: 10^{-7}/m/ year;

– safety valve fault (unexpected opening): 10^{-3}/year;

– pump fault: 10^{-1}/year;

– clogged pipes: 10^{-4}/year;

– electric fault of pump supply: 10^{-2}/year;

– regulation fault (sensor, regulator and actuator): 10^{-1}/year;

– fault of an "all or nothing" valve (unexpected opening or closing): 10^{-2}/year;

– first fault of an agitation system: 10^{-2}/year;

– electric supply breakdown: 10^{-2}/year;

– human loading error 10^{-3} per operation;

– PFD safety valve: 10^{-2};

– PFD of a All or Nothing (TOR) valve: 10^{-2};

– PFD of the sprinklers: 10^{-2};

– PFD cooling water (sprinklers): 10^{-2};

– PFD safety BPCS (sensor, regulator, actuator): 10^{-1};

– PFD alarm: 10^{-2}.

Appendix 8

XRisk Software

The XRisk software is a risk assessment tool. It manages the implementation of various methods, facilitates the representation of tables and graphs, provides assistance for the analysis of undesirable events, calculates probability risks and represents risk mapping. It is available at www.xrisk.fr.

XRisk works with the following methods: identification/simplified risk assessment, PHA, FMECA, HAZOP, SORAM, LOPA, bow-tie diagram, fault trees and causal trees.

All of these methods are simultaneously managed around a common internal model and the conversion of a representation into another one is automatic, including graphs. For example, an addition on the bow-tie diagram updates the FMECA table and vice versa. Moreover, the diagrams are automatically built from tables.

The systems analyzed can be modeled using a functional approach (IDEF0, SADT), a structural-functional approach or a process-oriented approach. These approaches allow us to describe the installation that is being analyzed. The model is built explicitly or implicitly once the risks are associated. It is built in an iterative manner.

The risk assessment can be carried out by using a qualitative, quantitative or mixed approach.

Bibliography

[AIC 01] AICHE, *Layer of Protection Analysis*, CCPS, 2001.

[ALE 02] ALE B., "Risk assessment practices in the Netherlands", *Safety Science*, vol. 40, pp. 105–126, 2002.

[ANN 67] ANNETT J., DUNCAN K., "Task analysis and training design", *Occupational Psychology*, vol. 41, pp. 211–221, 1967.

[ANN 00] ANNETT J., STANTON N., *Task Analysis*, Taylor & Francis, London, 2000.

[AVE 07] AVEN T., "A unified framework for risk and vulnerability analysis covering both safety and security", *Reliability Engineering and System Safety*, vol. 92, no. 6, pp. 745–754, 2007.

[AVE 10] AVEN T., *Misconception of Risk*, Wiley, London, 2010.

[AYY 01] AYYUB B., *Elicitation of Expert Opinions for Uncertainty and Risks*, CRC Press, NewYork, 2001.

[BAB 96] BABER C., STANTON N., "Human error identification techniques applied to public technology: predictions compared with observed use", *Applied Ergonomics*, vol. 27, no. 2, pp. 119–131, 1996.

[BEL 09] BELL J., HOLROYD J., Review of human reliability assessment methods, RR679, report, Health and Safety Executive, London, 2009.

[BIR 69] BIRNBAUM Z.W., *On the importance of different components in a multicomponent system*, KRISHNAIAH P.R. (ed.), *Multivariate Analysis II*, New York: Academic Press, pp. 581–592, 1969.

[BIR 96] BIRD F., GERMAIN G., *Practical Loss Control Leadership*, Det Norske Veritas Inc., Loganville, GA, 1996.

[BRA 03] BRANDENBUG H., WOJTYNA J., *L'approche processus Mode d'emploi*, Editions d'Organisation, Paris, 2003.

[CHA 06] CHARBONNIER J., *Le risk management: Méthodologie et pratiques*, Editions L'argus de l'assurance, Paris, 2006.

[COU 11] COUNCIL N. R., *Assessment of Approaches for Using Process Safety Metrics at the Blue Grass and Pueblo Chemical Agent Destruction Pilot Plants*, The National Academies Press, Washington, DC, 2011.

[COX 08] COX L., "What's wrong with risk matrices", *Risk Analysis*, vol. 28, no. 2, pp. 497–512, 2008.

[DAR 12] DARSA J.-D., *365 risques en entreprise: Une année en risk management*, Gereso, Le Mans, 2012.

[DEL 06] DELVOSALLE C., FIEVEZ C., PIPART A., *et al.*, "ARAMIS project: a comprehensive methodology for the identification of reference accident scenarios in process industries", *Journal of Hazardous Materials*, vol. 130, no. 3, pp. 200–219, 2006.

[DEN 98] DENIS H., *Comprendre et gérer les risques sociotechnologiques majeurs*, Ed. de l'Ecole Polytechnique de Montréal, Montréal, 1998.

[DEU 08] DEUST C., JOLY C., LENOBLE C., Estimation des aspects probabilistes-Guide pour l'intégration de la probabilité dans les études de dangers, report, DRA-08-95321-04393B, INERIS, 2008.

[DOD 12] DoD, "System safety program requirements MIL-STD-882", 2012.

[DOW 98] DOWELL A.M., "Layer of protection analysis for determining safety integrity level", *ISA Transactions*, vol. 37, pp. 155–165, 1998.

[DUG 92] DUGAN J., BAVUSO S., BOYD M., "Dynamic fault-tree models for fault-tolerant computer systems", *IEEE Transactions on Reliability*, vol. 41, no. 3, pp. 363–377, 1992.

[EMB 86] EMBREY D.E., "SHERPA: a systematic human error reduction and prediction approach", *International Topical Meeting on Advances in Human Factors in Nuclear Power Systems,* Knoxville, TN, pp. 184–193, 1986.

[ERI 97] ERIKSSON M., "A principal exposition of Jean-Louis Le Moigne's systemic theory", *Cybernetics and Human Knowing*, vol. 4, no. 2–3, pp.1–42, 1997.

[ERI 05] ERICSON C.A., *Hazard Analysis Techniques for System Safety*, Wiley, Hoboken, NJ, 2005.

[FLA 01] FLAUS J., "Une formalisation explicite de l'état et des flux dans la méthode MOSAR", *Congrès Français de Génie des Procédés*, Nancy, France, 17–19 October 2001.

[FLA 08] FLAUS J.M., "A model-based approach for systematic risk analysis", *Proceedings of the Institution of Mechanical Engineers, Part O: Journal of Risk and Reliability*, vol. 222, no. 1, pp. 79–83, 2008.

[FLE 75] FLEMING K.N., "A reliability model for common mode failures in redundant safety systems", *Proceedings of the 6th Annual Pittsburgh Conference on Modeling and Simulation,* General Atomic Report A 13284, 23–25, 1975.

[FRE 06] FREDERICKSON A., Layer of protection analysis, report, available at www.safetyusersgroup.com, May 2006.

[FUJ 04] FUJITA Y., HOLLNAGEL E., "Failures without errors: quantification of context in HRA", *Reliability Engineering and System Safety*, vol. 83, no. 2, pp. 145–151, 2004.

[FUM 01] FUMEY M., Méthode d'Evaluation des Risques Agrégés: application au choix des investissements de renouvellement d'installations, PhD Thesis, INPT, Industrial Systems, Toulouse, 2001.

[GAR 99] GARDES L., DEBRAY B., LONDICHE H., "Méthodologie d'analyse des risques dans les PME/PMI", *Qualita 99, 3e congrès international pluridisciplinaire qualité et sûreté de fonctionnement*, Besançon, France, 1999.

[GNE 12] GNESI S., MARGARIA T., *Formal Methods for Industrial Critical Systems: A Survey of Applications*, Wiley, Hoboken, NJ, 2012.

[GOB 10] GOBLE W., *Control Systems Safety Evaluation and Reliability, 3rd ed.*, Resources for measurement and control series, ISA. The Instrumentation, Systems, and Automation Society, Research Triangle Park, NC, 2010.

[GRA 05] GRAHAM P., COUNCIL I.R.G., *Risk governance: Towards an Integrative Approach*, International Risk Governance Council, http://www.irgc.org/IMG/pdf/IRGC_WP_No_1_Risk_Governance_reprinted_version_.pdf, Geneva, 2005.

[GRU 05] GRUHN P., CHEDDIE H., *Safety Instrumented Systems: Design, Analysis, and Justification*, ISA, The Instrumentation, Systems, and Automation Society, Research Triangle Park, NC, 2005.

[HAD 73] HADDON W., "Energy damage and the ten counter-measure strategies", *Human Factors Journal*, vol. 13, no. 4, pp. 321–331, 1973.

[HEL 04] HELTON J., OBERKAMPF W., "Alternative representations of epistemic uncertainty", *Reliability Engineering and Systems Safety*, vol. 85, 2004.

[HOL 93] HOLLNAGEL E., *Human Reliability Analysis: Context and Control*, Academic Press, New York, 1993.

[HOL 98] HOLLNAGEL E., *Cognitive Reliability and Error Analysis Method: CREAM*, Elsevier, Oxford, 1998.

[HOL 04] HOLLNAGEL E., *Barriers and Accident Prevention*, Ashgate, Aldershot, 2004.

[HOL 11] HOLLNAGEL E., PARIÈS J., WOODS D., *Resilience Engineering in Practice: A Guidebook*, Ashgate Studies in Resilience Engineering, Ashgate, 2011.

[HSE 78] HSE, The first canvey report, canvey: an investigation of potential hazards from operations in the canvey Island/ Thurrock Area, report, HMSO, London, 1978.

[HSE 01] HSE, Reducing risks, protecting people: HSE's decision making process, report, HMSO, London, 2001.

[ICS 09] ICSI, Fréquence des événements iniateurs d'accidents, report, *Les cahiers de la sécurité industrielle*, 2009.

[IEC 10a] IEC, "IEC 61508 functional safety of electrical/electronic/programmable electronic safety-related systems, edition 2 (2010), part 2", 2010.

[IEC 10b] IEC, "IEC 61508 functional safety of electrical/electronic/programmable electronic safety-related systems, edition 2 (2010), part 6, Annex D", 2010.

[INE 06] INERIS, Formalisation du savoir et des outils dans le domaine des risques majeurs - Omega 7, report, 2006.

[INE 08] INERIS, Evaluation des barrières techniques de sécurité - Omega 10, report, 2008.

[INE 09] INERIS, Démarche d'évaluation des Barrières Humaines de Sécurité - Omega 20, report, 2009.

[ISO 99] ISO, Guide ISO/CEI 51: safety aspects - guidelines for their inclusion in standards, 1999.

[ISO 09] ISO, ISO 31000:2009, risk management – principles and guidelines, 2009.

[JON 03] JONKMAN S., VAN GELDER P., VRIJLING J., "An overview of quantitative risk measures for loss of life and economic damage", *Journal of Hazardous Materials*, vol. 99, no. 1, pp. 1–30, 2003.

[KAP 81] KAPLAN S., GARRIC J., "On the quantitative definition of risk", *Risk Analysis*, vol. 1, pp. 11–27, 1981.

[KAR 10] KARAGIANNIS G., PIATYSZEK E., FLAUS J., "Industrial emergency planning modeling: a first step towards a robustness analysis", *Journal of Hazardous Materials*, vol. 181, pp. 324–334, 2010.

[KIR 97] KIRWAN B., KENNEDY R., TAYLOR-ADAMS S., *et al.*, "The validation of three human reliability quantification techniques, THERP, HEART and JHEDI: Part II - results of validation exercise.", *Applied Ergonomics*, vol. 28, no. 1, pp. 17–25, 1997.

[KLE 06] KLETZ T.A., *Hazop and Hazan*, Taylor & Francis, London, 2006.

[LAW 74] LAWLEY H.G., "Operability studies and hazard analysis", *Chemical Engineering Progress AIChE*, vol. 70, no. 4, 1974.

[LEM 94] LEMOIGNE J., *La théorie du système général: théorie de la modélisation*, PUF, Paris, 1994.

[LES 02] LESBATS M., CHAABANE S., DUTUIT Y., L'enseignement de la méthode "MADS-MOSAR" l'IUT Bordeaux 1, report, Journées Retour d'Expérience de la méthode d'analyse de risques MADS-MOSAR, Grenoble, 2002.

[LIN 06] LINDLEY D., *Understanding Uncertainty*, Wiley, Hoboken, NJ, 2006.

[MEE 10] MEEDDM, "Circulaire du 10/05/10 récapitulant les règles méthodologiques applicables aux études de dangers, l'appréciation de la démarche de réduction du risque la source et aux plans de prévention des risques technologiques (PPRT) dans les installations classées en application de la loi du 30 juillet 2003", *BO no. 2010/12*, 10 July, 2010.

[MEE 12] MEEDDM, Inventaire des accidents technologiques 2012, report, BARPI, 2012.

[NUR 75] NUREG/74/014, The WASH 400 report: reactor safety, an assessment of accident risks in US commercial nuclear power plant, report, US/NRC, 1975.

[NUR 07] NUREG/CR-6268, Common-cause failure database and analysis system, report, U.S. Nuclear Regulatory Commission, Washington, DC, 2007.

[OHS 09] OHSAS, OHSAS 18001 and 18002: Guidance for implementation of the OHSAS, report, ISO, London, 2009.

[PER 99] PERROW C., *Normal Accidents: Living with High-Risk Technologies*, Basic Books, New York, 1999.

[PER 07] PERILHON P., *La Gestion des risques: Méthode MADS-MOSAR II*, Editions Demos, Paris, 2007.

[POU 98] POUCET A., Human factors reliability benchmark exercise, report, European Commission, 1998.

[RAS 83] RASMUSSEN J., "Skills, rules, knowledge: signals, signs and symbols and other distinctions in human performance models", *IEEE Transactions on Systems, Man and Cybernetics*, vol. 13, pp. 257–267, 1983.

[REA 90] REASON J., *Human Error*, Cambridge University Press, Cambridge, 1990.

[REA 97] REASON J., *Managing the risks of organizational Accidents*, Ashgate, 1997.

[REI 99] REID S., "Perception and communication of risk, and the importance of dependability", *Structural Safety*, vol. 21, no. 4, pp. 373–384, 1999.

[REN 92] RENN O., BURNS W.J., KASPERSON J.X., *et al.*, "The social amplification of risk: theoretical foundations and empirical applications", *Journal of Social Issues*, vol. 48, no. 4, pp. 137–160, 1992.

[REN 98] RENN O., "The role of risk perception for risk management", *Reliability Engineering and System Safety*, vol. 59, no. 1, pp. 49–62, 1998.

[ROB 90] ROBERTS K.H., "Some characteristics of one type of high reliability organization", *Organization Science*, vol. 1, no. 2, 1990.

[ROS 77] ROSS D.T., "Structured analysis: a language for communicating ideas", *IEEE Transactions on Software Engineering*, vol. SE 3:1, 1977.

[SAL 03] SALMON P., STANTON N., YOUNG M., *et al.*, "Predicting design induced pilot error: a comparison of SHERPA, human error HAZOP, HEIST and HET, a newly developed aviation specific HEI method", *Human-Centred Computing–Cognitive, Social and Ergonomic Aspects,* Lawrence Erlbaum Associates, London, 2003.

[SAL 06] SALVI O., DEBRAY B., "A global view on ARAMIS, a risk assessment methodology for industries in the framework of the SEVESO II directive", *Journal of Hazardous Materials*, vol. 130, pp. 187–199, 2006.

[SAW 87] SAWIN A., Accident sequence evaluation program (ASEP), report, NUREG/CR-4772, US Nuclear Regulatory Commission, Washington, DC, 1987.

[SCH 01] SCHABE H., "Different principles used for determination of tolerable hazard rates", *Materials of the World Congress on Railway Research,* Cologne, Germany, 2001.

[SCH 09] SCHUBERT U., DIJKSTRA J.J., "Working safely with foreign contractors and personnel", *Safety Science*, vol. 47, no. 6, pp. 786–793, 2009.

[SHA 07] SHAPPELL S., DETWILER C., HOLCOMB K., *et al.*, "Human error and commercial aviation accidents: an analysis using the human factors analysis and classification system", *Human Factors: The Journal of the Human Factors and Ergonomics Society*, vol. 49, no. 2, pp. 227–242, April 2007.

[SHO 12] SHOUHED D., GEWERTZ B., WIEGMANN D., et al., "Integrating human factors research and surgery: a review", *Archives of Surgery*, vol. 147, no. 12, pp. 1141–1146, 2012.

[SIM 96] SIMON H.A., *The Sciences of the Artificial*, MIT Press, Cambridge, MA, 1996.

[SKL 06a] SKLET S., "Safety barriers: definition, classification, and performance", *Journal of Loss Prevention in the Process Industries*, vol. 19, no. 5, pp. 494–506, 2006.

[SKL 06b] SKLET S., "Hydrocarbon releases on oil and gas production platforms: release scenarios and safety barriers", *Journal of Loss Prevention in the Process Industries*, vol. 19, no. 5, pp. 481–493, 2006.

[SLO 01] SLOVIC P., "The risk game", *Journal of Hazardous Materials*, vol. 86, no. 1–3, pp. 17–24, 2001.

[SMI 00] SMITH D., *Reliability, Maintainability and Risk*, Elsevier Science, London, 2000.

[SOM 10] SOMMERVILLE I., *Software Engineering*, 9th ed., Pearson, Redwood City, CA, 2010.

[STA 99] STANTON N.A., YOUNG M., *Guide to Methodology in Ergonomics: Designing for Human Use*, Taylor & Francis, London, 1999.

[SWA 83] SWAIN A., GUTTMANN H., Handbook of human reliability analysis with emphasis on nuclear power plant applications, report NUREG/CR-1278, USNRC, 1983.

[SWA 87] SWAIN A., Accident sequence evaluation program: human reliability analysis procedure, report NUREG/CR-4772, Nuclear Regulatory Commission, Washington, DC, 1987.

[VIL 97] VILLEMEUR A., *Sûreté de fonctionnement des systèmes industriels: fiabilité - facteurs humains - informatisation*, Collection de la Direction des Etudes et Recherches d'Electricité de France, Eyrolles, 1997.

[WEI 39] WEIBULL W., "A statistical theory of the strength of materials", Ingeniörsvetenskapsakademiens handlingar, Generalstabens litografiska anstalts förlag, http://www.barringer1.com/wa_files/Weibull-1939-Strength-of-Materials.pdf, 1939.

[WEI 08] WEILKIENS T., *Systems Engineering with SysML/UML*, Elsevier Science, London, 2008.

[WIL 58] WILLIAMS H., "Reliability evaluation of the human component in man-machine systems", *Electrical Manufacturing*, pp. 78–82, 1958.

[WIL 85] WILLIAMS J., "HEART - A proposed method for achieving high reliability in process operation by means of human factors engineering technology", *Proceedings of the Symposium on the Achievement of Reliability in Operating Plant, Safety and Reliability Society*, Southport, NC, 1985.

[WOO 94] WOODS D., JOHANNESEN L., COOK R., SARTER N., Behind human error: cognitive systems, computers and hindsight, report, CSERIAC, The Ohio State University, 1994.

Index

β-factor, 250, 251

A

acceptability, 112, 113, 117
accident, 20, 273, 274, 343, 344
  scenario, 19
activity diagram, 139, 140
ALARA, 29–31
AND connector, 80
ANSI/ISA, 303
AS/NZS, 53
ASEP, 266, 282
axiomatic approach, 97

B

Bayesian approach, 97
  network, 84–87
bow-tie-type diagram, 83, 84, 259–262

C

causal tree, 79, 80
cause of a failure, 181–183
cognitive behaviour, 271
  model, 274

commission error, 272, 273
common cause, 197, 247–251
CPC, 290
CREAM, 288–291
criticality, 23, 109, 110, 186, 187

D

decision/action flow diagram, 144
deductive approach, 26, 27
definitions, 156–158, 181, 231, 232, 265–267, 335, 336
detection of a failure, 185, 186
deviation, 206–208
dynamic fault tree, 89, 90

E

effect of a failure, 183, 184
EPC, 281
event tree, 82, 144, 145, 253–259
  sequence, 255
expert judgment, 101

F

failure, modes, effects and criticality analysis (FMECA), 209, 226

fault rate, 240
　tree, 80, 81
final event, 19, 316
FMEA, 197, 198
frequentist approach, 97, 98
functional modelling, 128–131, 134–137

H

hazard, 16, 17, 158–163
hazardous event, 19, 158, 159
　situation, 159–163
HAZOP, 201–208
HEART, 280–282
hierarchical breakdown, 128, 129
　task analysis (HTA), 141–144
human barrier, 223
　error, 269–278

I

IDEF0, 131, 138, 349–351
IEC 60601, 303
　60812, 180
　61508, 303, 304
　61511, 303
　61513, 303
　62061, 303
　62278, 303
immediate, necessary and sufficient (INS) cause, 231
incident, 20
independent protection layer (IPL), 308, 309, 317
inductive approach, 26, 27
insignificant, 117
intensity, 22
Ishikawa, 78
ISO 26262, 303
　31000, 55–61

K, L

keywords, 205
lapses, 272
likelihood, 21, 22, 96–102, 163
LOPA, 307–318

M

machine-like behaviour, 270
major industrial risk, 39, 40
Markov, 87–89
material barrier, 296
MIL-STD-882, 105, 111
minimal cut sets, 237–241, 247
MOCUS, 238, 239
MTBF, 335

O

OAET, 144
omission error, 272, 273
operator activity, 140
OR connector, 80

P

PFH, 306,
preliminary hazard analysis, 165–169
prevention, 23–25
probability of
　failure on demand (PFD), 315, 360
　human error, 266
procedural behaviour, 270
process instrumentation
　diagram, 202
professional risk, 341
protection, 23–25
PSC, 231
PSF, 276

R

reason, 48–51, 75–77, 271, 272
reliability diagram, 242, 243
resilience, 18
risk
   management process, 61–69
   matrix, 110–112, 116–118, 120–122
RPN, 186, 187

S

SADT, 131, 349–351
safety, 20, 21, 301–307, 315, 316
   instrumented system (SIS), 301–307
   integrity level (SIL), 300
severity, 22, 102–109, 116, 118–120, 163, 185, 194, 222
   of a failure, 185
SHERPA, 278–280
slips, 271
source of hazard, 16, 17, 215–220
SRK, 270, 271
stake, 17, 18
structural modelling, 131–134
structural-functioning, 352–354
subjective approach, 98
subsystems, 127
systematic and organized risk analysis method (SORAM), 211, 212
   table A, 216, 220
   table B, 224, 227

T

target, 17, 18, 35, 36, 74, 75
temporal frequency, 99, 100
THERP, 282–288
trigger event, 160

U

UIC, 167
uncertainty, 93

V, W

violations, 50, 272
vulnerability, 18
Weibull, 333